Lecture Notes in Networks and Systems

Volume 38

Series editor

Janusz Kacprzyk, Polish Academy of Sciences, Warsaw, Poland
e-mail: kacprzyk@ibspan.waw.pl

The series "Lecture Notes in Networks and Systems" publishes the latest developments in Networks and Systems—quickly, informally and with high quality. Original research reported in proceedings and post-proceedings represents the core of LNNS.

Volumes published in LNNS embrace all aspects and subfields of, as well as new challenges in, Networks and Systems.

The series contains proceedings and edited volumes in systems and networks, spanning the areas of Cyber-Physical Systems, Autonomous Systems, Sensor Networks, Control Systems, Energy Systems, Automotive Systems, Biological Systems, Vehicular Networking and Connected Vehicles, Aerospace Systems, Automation, Manufacturing, Smart Grids, Nonlinear Systems, Power Systems, Robotics, Social Systems, Economic Systems and other. Of particular value to both the contributors and the readership are the short publication timeframe and the world-wide distribution and exposure which enable both a wide and rapid dissemination of research output.

The series covers the theory, applications, and perspectives on the state of the art and future developments relevant to systems and networks, decision making, control, complex processes and related areas, as embedded in the fields of interdisciplinary and applied sciences, engineering, computer science, physics, economics, social, and life sciences, as well as the paradigms and methodologies behind them.

Advisory Board

Fernando Gomide, Department of Computer Engineering and Automation—DCA, School of Electrical and Computer Engineering—FEEC, University of Campinas—UNICAMP, São Paulo, Brazil
e-mail: gomide@dca.fee.unicamp.br

Okyay Kaynak, Department of Electrical and Electronic Engineering, Bogazici University, Istanbul, Turkey
e-mail: okyay.kaynak@boun.edu.tr

Derong Liu, Department of Electrical and Computer Engineering, University of Illinois at Chicago, Chicago, USA and Institute of Automation, Chinese Academy of Sciences, Beijing, China
e-mail: derong@uic.edu

Witold Pedrycz, Department of Electrical and Computer Engineering, University of Alberta, Alberta, Canada and Systems Research Institute, Polish Academy of Sciences, Warsaw, Poland
e-mail: wpedrycz@ualberta.ca

Marios M. Polycarpou, KIOS Research Center for Intelligent Systems and Networks, Department of Electrical and Computer Engineering, University of Cyprus, Nicosia, Cyprus
e-mail: mpolycar@ucy.ac.cy

Imre J. Rudas, Óbuda University, Budapest, Hungary
e-mail: rudas@uni-obuda.hu

Jun Wang, Department of Computer Science, City University of Hong Kong Kowloon, Hong Kong
e-mail: jwang.cs@cityu.edu.hk

More information about this series at http://www.springer.com/series/15179

Mohan L. Kolhe · Munesh C. Trivedi
Shailesh Tiwari · Vikash Kumar Singh
Editors

Advances in Data and Information Sciences

Proceedings of ICDIS-2017, Volume 1

 Springer

Editors
Mohan L. Kolhe
Smart Grid and Renewable Energy
University of Agder
Kristiansand
Norway

Munesh C. Trivedi
Department of Computer Science
    and Engineering
ABES Engineering College
Ghaziabad, Uttar Pradesh
India

Shailesh Tiwari
Department of Computer Science and
    Engineering
ABES Engineering College
Ghaziabad, Uttar Pradesh
India

Vikash Kumar Singh
Department of Computer Science
    and Engineering
The Indira Gandhi National
    Tribal University
Amarkantak, Madhya Pradesh
India

ISSN 2367-3370   ISSN 2367-3389   (electronic)
Lecture Notes in Networks and Systems
ISBN 978-981-10-8359-4   ISBN 978-981-10-8360-0   (eBook)
https://doi.org/10.1007/978-981-10-8360-0

Library of Congress Control Number: 2018933483

Printed on acid-free paper

This Springer imprint is published by the registered company Springer Nature Singapore Pte Ltd. part of Springer Nature
The registered company address is: 152 Beach Road, #21-01/04 Gateway East, Singapore 189721, Singapore

Preface

ICDIS-2017 is a major multidisciplinary conference organized with the objective of bringing together researchers, developers, and practitioners from academia and industry working in all areas of computer and computational sciences. It is organized specifically to help computer industry to derive the advances in next-generation computer and communication technology. Researchers invited to speak will present the latest developments and technical solutions.

Technological developments all over the world are dependent upon the globalization of various research activities. Exchange of information and innovative ideas is necessary to accelerate the development of technology. Keeping this ideology in preference, the International Conference on Data and Information Sciences (ICDIS-2017) has been organized at Indira Gandhi National Tribal University, Amarkantak, MP, India, during November 14–15, 2017.

The International Conference on Data and Information Sciences has been organized with a foreseen objective of enhancing the research activities at a large scale. Technical Program Committee and Advisory Board of ICDIS-2017 include eminent academicians, researchers, and practitioners from abroad as well as from all over the nation.

In this book, selected manuscripts have been subdivided into two tracks namely— smart hardware and software design and smart computing techniques. A sincere effort has been made to make it an immense source of knowledge by including 27 manuscripts in this proceedings volume. The selected manuscripts have gone through a rigorous review process and are revised by authors after incorporating the suggestions of the reviewers. These manuscripts have been presented at ICDIS-2017 in different technical sessions.

ICDIS-2017 received around 230 submissions from around 550 authors of different countries such as India, Malaysia, Bangladesh, Sri Lanka. Each submission went through the plagiarism check. On the basis of plagiarism report, each submission was rigorously reviewed by at least two reviewers with an average of 1.93 per reviewer. Even some submissions had more than two reviews. On the basis of these reviews, 59 high-quality papers were selected for publication in two proceedings volumes, with an acceptance rate of 25.6%.

We are thankful to our keynote speakers, delegates, authors for their participation and interest in ICDIS-2017 as a platform to share their ideas and insights. We are also thankful to Prof. Dr. Janusz Kacprzyk, Series Editor, AISC, Springer Nature, and Mr. Aninda Bose, Senior Editor, Hard Sciences, Springer Nature, India, for providing continuous guidance and support. Also, we extend our heartfelt gratitude to the reviewers and Technical Program Committee members for showing their concern and efforts in the review process. We are indeed thankful to everyone directly or indirectly associated with the conference organizing team leading it toward the success.

Although utmost care has been taken in compilation and editing, a few errors may still occur. We request the participants to bear with such errors and lapses (if any). We wish you all the best.

Organizing Committee
ICDIS-2017

Norway   Mohan L. Kolhe
Ghaziabad, UP, India   Shailesh Tiwari
Ghaziabad, UP, India   Munesh C. Trivedi
Amarkantak, MP, India   Vikash Kumar Singh

Organizing Committee

Patron	Prof. T. V. Kattimani, Vice-Chancellor, Indira Gandhi National Tribal University, Amarkantak, India
General Chairs	Dr. Mohan L. Kolhe, University of Agder, Norway
	Dr. Shekhar Pradhan, DeVry University, New York
Program Chair	Dr. K. K. Mishra, Motilal Nehru National Institute of Technology Allahabad, India
	Prof. Shailesh Tiwari, ABES Engineering College, Ghaziabad, UP, India
Conference Chairs	Dr. Vikas Kumar Singh, Indira Gandhi National Tribal University, Amarkantak, MP, India
	Dr. Munesh C. Trivedi, ABES Engineering College, Ghaziabad, UP, India
TPC Chairs	Dr. Nitin Singh, Motilal Nehru National Institute of Technology Allahabad, India
	Dr. B. K. Singh, RBS College, Agra, UP, India
Publication Chairs	Dr. Deepak Kumar Singh, Sachdeva Institute of Technology, Mathura, India
	Dr. Pragya Dwivedi, Motilal Nehru National Institute of Technology Allahabad, India
Publicity Chairs	Dr. Anil Dubey, Government Engineering College, Ajmer, India
	Dr. Deepak Kumar, Amity University, Noida, India
	Dr. Nitin Rakesh, Amity University, Noida, India
	Dr. Ravi Prasad Valluru, Narayana Engineering College, Nellore, AP, India

Dr. Sushant Upadyaya, MNIT, Jaipur, India
Dr. Akshay Girdhar, GNDEC, Ludhiana, India

Publicity Co-Chair   Prof. Vivek Kumar, DCTM, Haryana, India

About the Book

With the advent of the digital world, the information and data science came into existence with a wide scope of innovations and implementations. Both of these play a major role in the making of policies and taking decisions within or outside any organization, institution, society, etc.

Data science and information science are complementary to each other, but distinct. Data science is related to an inference of knowledge and meaningful information from data. However, information science deals with the design and development of strategies, methods, and techniques concerned with the analysis, classification, storage, retrieval, dissemination, and protection of information.

Nowadays, information and data science field entered into a new era of technological advancement, which we call *Smart and Intelligent Information and Data Science*. *Smart and Intelligent Information and Data Science* provides the use of artificial intelligence techniques to solve the complex problems related to policy and decision making. We can say that the main objective of *Ambient Computing and Communication Sciences* is to make software, techniques, computing and communication devices, which can be used effectively and efficiently.

Keeping this ideology in preference, this book includes the insights that reflect the immediate surroundings developments in the field of *Smart and Intelligent Information and Data Science* from upcoming researchers and leading academicians across the globe. It contains the high-quality, peer-reviewed papers of '*International Conference on Data and Information Sciences (ICDIS-2017)*', held at Indira Gandhi National Tribal University, Amarkantak, MP, India, during November 17–18, 2017. These papers are arranged in the form of chapters. The contents of this book cover two areas: *Smart Hardware and Software Design, Smart Computing Techniques*. This book helps the prospective readers from industry and academia to derive the immediate surroundings developments in the field of data and information sciences and shape them into real-life applications.

Contents

About the Editors

Prof. (Dr.) Mohan L. Kolhe is with the University of Agder, Norway, as Full Professor in Electrical Power Engineering with focus on smart grid and renewable energy in the Faculty of Engineering and Science. He has also received the offer of a full professorship in the smart grid from the Norwegian University of Science and Technology (NTNU). He has more than 25 years' academic experience at the international level on electrical and renewable energy systems. He is a leading renewable energy technologist and has previously held academic positions at the world's prestigious universities, e.g., University College London, UK/Australia; University of Dundee, UK; University of Jyvaskyla, Finland; Hydrogen Research Institute, QC, Canada.

Prof. (Dr.) Munesh C. Trivedi currently works as a Professor in Computer Science and Engineering Department, ABES Engineering College, Ghaziabad, India. He has published 20 textbooks and 80 research publications in different international journals and proceedings of international conferences of repute. He has received Young Scientist and numerous awards from different national as well as international forums. He has organized several international conferences technically sponsored by IEEE, ACM, and Springer. He is in the review panel of IEEE Computer Society, International Journal of Network Security, Pattern Recognition Letters, and Computers & Education (Elsevier's Journal). He is Executive Committee Member of IEEE UP Section, IEEE India Council, and also IEEE Asia Pacific Region 10.

Prof. (Dr.) Shailesh Tiwari currently works as a Professor in Computer Science and Engineering Department, ABES Engineering College, Ghaziabad, India. He is an alumnus of Motilal Nehru National Institute of Technology Allahabad, India. His primary areas of research are software testing and implementation of optimization algorithms and machine learning techniques in various problems. He has published more than 50 publications in international journals and in proceedings of international conferences of repute. He is editing Scopus, SCI, and E-SCI-indexed journals. He has organized several international conferences under the banner of IEEE and Springer. He is a Senior Member of IEEE, Member of IEEE Computer Society, Fellow of Institution of Engineers (FIE).

Dr. Vikash Kumar Singh is with Indira Gandhi National Tribal University, Amarkantak, MP, India, as Associate Professor in Computer Science with focus on artificial intelligence in the Faculty of Computronics. He has also received UGC-NET/JRF. He has more than 17 years of academic experience. He has completed MCA along with Ph.D. His academic and research work includes more than 250 research papers, and he has attended more than 15 national and international conferences, workshops, and seminars. He has been invited by many national/international organizations for delivering expert lectures/courses/keynote addresses/workshops.

Part I
Smart Hardware and Software Design

Computation of Dynamic Signal Phases for Vehicular Traffic

Rajendra S. Parmar and Bhushan H. Trivedi

Abstract Traffic congestion is one of the significant contributors to global warming and a major factor deteriorating logistic efficiency, thereby impeding efficiencies and economy. Infrastructure expansion and improvements render a short-lived solution. Significant improvisation is achieved only through technological approach to the problems with challenges in signaling shortest route and ensuring fair treatment to all the directions without a deadlock-like situation. One of the firsts to address is signaling electronics. Signaling electronics are not maintained. Besides, traffic signals do not align with the traffic patterns. Traffic congestion is characterized as dynamic, stochastic, random, and unpredictable phenomenon. Consequently, traffic congestion cannot be addressed by static, predetermined signal phases, preprogrammed periodically changing signals based on a prior knowledge of traffic behavior as the dynamically changing traffic pattern will dislocate and disrupt the assumptions of traffic changes. Hence traffic signals have to be devised to adapt to changes as per the traffic situation. Secondly, signaling electronics is oblivious to vehicle densities, intended direction of travel, and available capacity on road ahead to accommodate oncoming vehicles. This leads to a green phase resulting in deadlock situation. The paper describes exploring utilization of green phases for other directions while maintaining exclusive and non-conflicting movement with other directions. The paper proposes computation of dynamic signal phases while ensuring fair assignment to other directions and avoidance of deadlock situations.

Keywords Traffic signals · Intelligent traffic signals · Dynamic traffic signal assignment

R. S. Parmar (✉)
GTU, Gujarat Technology University, Ahmedabad 382424, India
e-mail: raj@emerging-india.com

B. H. Trivedi
GLS Institute of Computer Technology, Ahmedabad 380006, Gujarat, India
e-mail: bhtrivedi@gmail.com

© Springer Nature Singapore Pte Ltd. 2018
M. L. Kolhe et al. (eds.), *Advances in Data and Information Sciences*, Lecture Notes in Networks and Systems 38, https://doi.org/10.1007/978-981-10-8360-0_1

1 Introduction

1.1 Challenges

The signal phase definition has several challenges:

a. Traffic signals are randomly spaced; there is no orderly positioning. Locations of traffic signals are based on commuter convenience rather than traffic considerations. As a result, signaling islands are randomly located. Distance between consecutive traffic signal is random.
b. Road geometry or town planning has not considered having fixed number of converging roads at junctions. Meaning a junction with four roads can lead to another junction with three diverging roads.
c. Infrastructure providers often fail to provide uniform road quality, road width or number of lanes. This too leads to reducing traffic speeds and giving rise to congestion.
d. Different sizes of vehicles have different footprints in width as well as length. A wider vehicle leaves plenty of empty space around it which results in loss of usable area. If the vehicles were relatively of similar size, more vehicles would have been accommodated.
e. Unruly drivers and pedestrians are mostly responsible for incidents of traffic congestion. World over slow driving is appreciated; however, a slower driver often is responsible for congestion situation.

1.2 Objectives, Methods, Strategies

a. Objective
   The basic objective is (1) scheduling for different time of the day (2) dynamic signaling based on traffic patterns (3) optimization methods [1, 2].
   The overall objectives of traffic signaling are to:

   1. Achieve disciplined and systematic movement of vehicle to ensure smooth traffic flow.
   2. Reduce delays by reducing stoppages, accidents, crashes, and severity frequency.
   3. Quantitative measurements of traffic signaling parameters.

b. Methods
   Traffic control methods are classified [3] as:

   1. Timing control: A central computer helps in gaining control over this function. However, should the central computer fail, the entire operation comes to a halt.

2. Half-induced control: Signal controller and region controller control traffic through coordination. Control strategy is defined; however, it has inherent latency.
3. Complete-induced control: No communication between intersections, hence not a best method.
4. Adaptive control: Lack of communication between intersections results in lack of proactive information to the driver, hence avoidable method.

c. Strategy
Some of the strategies are:

1. Fixed time traffic signal: Based on past traffic data, signal phases are preprogrammed. Now obsolete.
2. AI and fuzzy logic approach: Extension of green phase is decided by neural network.
3. V2V (vehicle-to-vehicle) and V2I (vehicle-to-infrastructure) communication are suggested for traffic signaling and subsequent congestion mitigation.
4. Genetic algorithm: Derived from swarm intelligence. Learning is only the fittest survive; most surviving traffic signal phase is green. Genetic algorithm is also used for rerouting.
5. Reinforcement learning: Learning from environment to exploit most suitable reaction to dynamic situations. Deployed to manage traffic queue. Q-Learning is a reinforcement learning applied in optimization of the traffic flow.

1.3 Desired Solution

The following is a feature list of a desired solution:

a. The signal phase assignment must be fair to all the direction. Meaning even if a direction has constant incoming traffic, it cannot have unreasonably long green phase assignment. Secondly, every direction must have a signal phase assignment in every signal phase cycle.
b. Boundary conditions must be defined to specify minimum and maximum time for every signal phase. This will ensure against unrealistic duration of signal phases and can possibly even achieve deadlock avoidance.
c. Number of vehicle in amalgamation with occupied footprint should be considered rather than only the number of vehicles as a criterion to assign signal phases.
d. Unutilized signal phases must be utilized such that vehicles can move to other non-conflicting directions for the assign signal phase.

2   Prior Research—Signal Phases

2.1   Modeling

Due to the probabilistic nature of vehicular traffic, it draws similarities from fluid dynamics and vehicular traffic, which are represented by analytical models, hence analytical models are studied. Lighthill, Whitham, Richard continuum model widely known as LWR model, is proposed [4]. The model relates traffic density and vehicular speed. Indirectly based on flow, the model suggests preferred speeds for vehicles. The first-order relationship is expressed as follows:

$$\frac{dk\,(x,t)}{dt} + \frac{dq\,(x,t)}{dx} = 0$$

where k is density; q is vehicular flow; both these parameters are function of time and position.

However, different vehicles have different sizes and different speeds. Also, the start and stops are so abrupt that the continuum model cannot address challenges of real-world vehicular traffic.

Lacuna of LWR model is addressed by introducing additional terms in the equation [5]. These modifications expect vehicles to align their speeds based on vehicles at rear; like in fluid dynamics. However for vehicular traffic, speeds are governed by vehicles in the front. The analytical model helps appreciate congestion as a spatialtemporal problem in time and space coordinates. Vehicular traffic congestion has several equilibrium states. Higher-order model introduces disproportionately higher complexities which add humongous computational burden.

2.2   Solution Approaches

a. Petri Nets

Paper Demonstrated exploiting Petri Nets for representing vehicular traffic elements. Petri Net representation help model events which are isolated, concurrent, conditional or synchronized inclusive of constraints [6], precedential conditions, and frequency of occurrence. Petri Nets are quick to coding. Petri Net representation is shown in Fig. 1.

The paper incorporates coordination of adjacent traffic signals in assigning signal phases [7]. Petri Nets are deployed with time intervals; both endpoints minimum and maximum are associated with places. It delivers responsive traffic signal timing as it extends green time based on demand.

b. Queuing

Fig. 1 Petri Net representation. *GG* Go green; *EG* extend green; *Act* transition actuation; *FO* force off; *RG* rest green; *DG* display green; *M* movement; *GR* go red

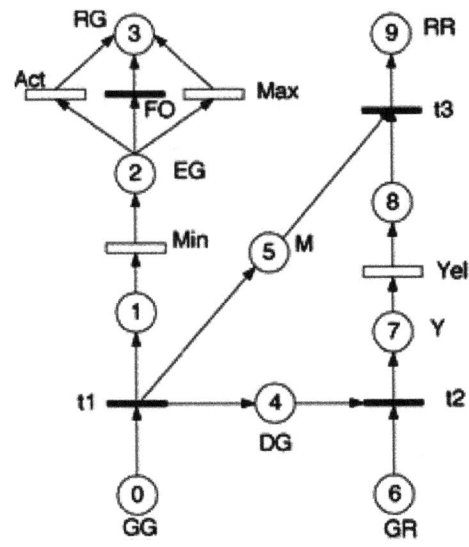

Any timing signal cannot sustain increasing traffic with limited infrastructure [8]. Traffic forms a queue and hence queue learning is important. Algorithm works on iterative methods where improvements are implemented, and results are studied in an iterative process. Queue is a FIFO [9]. The paper exploits $M/M/1$ queuing system which is a widely used model of queuing system. The queue dispensing must always be higher than queue building. It addresses traffic congestion to reduce waiting time of vehicles by effective signaling phases.

c. Modeling; Statistical Approach

Proposes a simplified approach by defining a minimum cycle time [1].

$$C = (1.5 * L + 5) / (1 - Y)$$

C = Cycle length in seconds; L = Lost time or [(no. of phases) × (lost time per phase) + (all red time)]; Y = Critical lane volume divided by the saturation flow, summed over the phases, where saturation flow is equivalent per hourly rate at which vehicles can traverse an intersection approach under prevailing conditions, assuming a constant green indication at all time and no loss time, in vehicles per hour or vehicles per hour per lane. Poisson distribution is also employed to address traffic congestion. It suffers from the limitations of statistical methods which are empirical, applicable for large data sets and are prone to errors when unprecedented situation occurs. Vehicle movement is nonlinear, dynamic and stopping intermittently over 2D space [10]. This is then extended to area with finite congestion and the approaches are enumerated in detail [11].

d. Wireless Sensor Network (WSN)

WSN is used to sense vehicles entering and leaving a road segment [12]. The demand from the direction with highest number of vehicles is assigned green phase. A low-power, low-cost WSN installed on roadside gathers information and controls traffic lights [13]. Three WSN sensors help vehicles to compute its speed. WSN can then decide if the vehicle can cross the junction before the expiry of signal phase or not. WSN network senses congestion [14, 15]. The model is based on VISSIM platform, currently, limited to single traffic junction.

e. Adhoc Networks

Consider statically changing traffic lights and advises rerouting to achieve shortest path [16]. The system is called Shortest-Path-Based Traffic-Light-Aware Routing (STAR). It deploys ad hoc Networks like MANET, VANET.

f. Coordinated Signals

Paper proposes to synchronize traffic lights and advises travel speeds to commuters such that vehicles do not have to stop [17–19]. However, maintaining travel speed is difficult for several reasons; primarily, errors in speed measurements and drivers ignoring the advice. Mobile agents communicate between line sensors, segment sensors, vehicle sensors, signal controllers, and video cameras. The mobile agents collect data from sensors, analyze for problems in traffic and communicate with roadside cameras which calculate signal timing.

g. Self Learning Traffic Signals

Present traffic signals which have static timing and do not have flexibility to accept dynamic change in signal timing [8]. AI and fuzzy logic are proposed to address dynamically changing traffic behavior. It proposes a continuously learning system to change the traffic plans; however, systems fail when unprecedented situation arises, following which it has to accept forced learning or behave according to a small set of pre-fed data.

h. Other Approaches

Paper proposes that the only way to resolve traffic congestion is to provide real-time information to road users so that they can make informed decisions on speed, route, and timing [20]. It uses vehicles traveling in both the directions to get traffic information and requires no dedicated roadside infrastructure. The wireless technology is put to use to get information, and a signboard is created to offer information which can be referred by the road users. Paper develops real-time signal timing according to actual traffic flow [21]. The model decides signal timing to relieve traffic congestion. Release matrix is employed to address randomness of traffic situations. Release matrix analyzes traffic situations and minimizes total delay by using maximum road capacity. Paper prosposes a vision-based closed-loop system [22]. However, disadvantages of vision-based systems continue to challenge accuracy and repeatability. Ant colony optimization (ACO) offers distributed intelligent traffic system (DITS) [23]. Intelligence collected by vehicles is distributed and hence vehicles do not travel congested roads; very akin to ACO which does not travel where pheromone is not present. For intelligence to be uniformly available, a centralized information system is required. Probe vehicles are employed to get real-time data to compile traffic condition [24]. The approach provides good results with larger number of probe vehicles.

Fig. 2 Signal direction

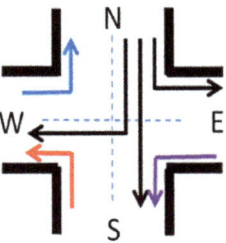

2.3 Contemporary Systems

It collates information about vehicle position through crowd-sourcing; from map users with Android OS, excluding iPhones [25]. The data acquired helps compute speed through indirect measurements. The satellite-based component to integrate with GPS is Wide Area Augmentation System (WAAS) in USA, European Geostationary Navigation Overlay System EGNOS in Europe, Multi-Functional Satellite Augmentation System MSAS in Japan. Contemporary systems use Windows-based OS in collaboration with traffic controllers from different manufacturers. Split Cycle Offset Optimization Technique (SCOOT) is deployed to control the signal phases in real time [26]. It collates information from sensors installed at several positions in the entire infrastructure. Sydney Coordinated Adaptive System (SCAT) is an Automatic Traffic Control System (ATCS) which generates signal phases using complex algorithms [27]. Here too, the data is collected from infrastructure sensors. Other significant development is Urban Traffic Optimization by Integrated Automation (UTOPIA), which considers vehicles detection, vehicle flow, traffic intensity, road occupancy, and even type of vehicle. Federal Highway Administration (FHWA) developed ACS Lite is widely deployable at much reduced cost [28]. RHODES is ATCS technology which predicts traffic levels from inductive loop sensors [29]. The paper compares available sensor technology and concludes wireless sensors with near nil cost are the best alternative [30, 31].

3 Dynamic Signal Phases

3.1 Maximizing Signal Phase Opportunity

Many times, assigned green phase is not utilized as there are no waiting vehicles [32]. During such instances, evaluate if there exist other non-conflicting, green phase-hungry directions that can be served. This is illustrated in Figs. 2, 3 and 4.

Traffic Schema (Right-Hand Drive)
In case there is not NS traffic, the signal phase is utilized for SE or EW as shown in Fig. 3.

Fig. 3 Absence of NS traffic

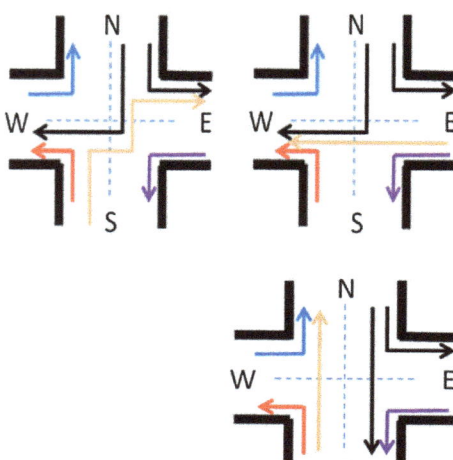

Fig. 4 Absence of NW traffic

Option 1 of utilizing the absence of SN traffic (Note—This discussion is based on right-hand drive vehicles) (Fig. 4).

If NW is absent, it can be handed over to SN.

3.2 Vehicle Flow

It is essential to establish a relationship between inflow and outflow of vehicles at a traffic junction to ensure every green phase assignment would at least evacuate vehicles and not accumulate more than present during assignment of green phase. At any time, cumulative outflow must be higher than cumulative inflow else congestion situation shall prevail. Duration of signal phase must be proportional to a combination of vehicle count, inflow rate, and ARCA. A simulation study is undertaken where different outflow, inflow, and vehicle count at the beginning can be defined. For the simulation, a fixed inflow rate of 600 vehicles/hour is considered and different outflow rate from 1,200/1,800/2,000/2,400 vehicles/hour is plotted in Fig. 5.

3.3 Core Engine

The hypothesis of fair assignment of dynamic signal phases is based on road capacity, road occupancy, vehicle count, and cumulative waiting time. The most important parameter is evacuation speed which is speed of travel on the road segment ahead.

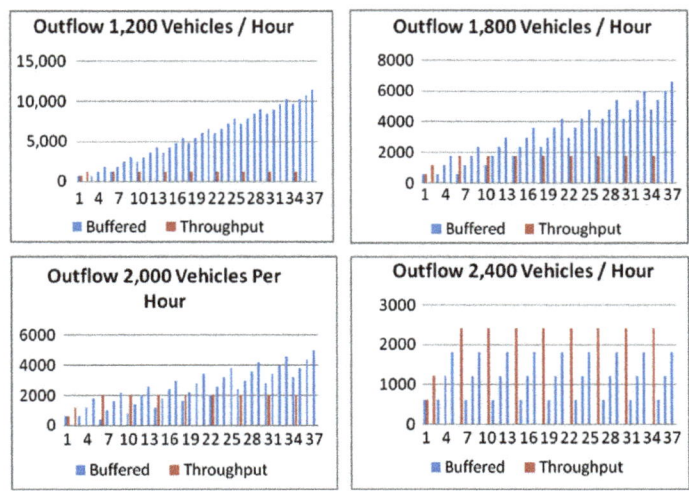

Fig. 5 Vehicle inflow vis-à-vis outflow

1. A vehicle waiting maximum time must get priority.
2. However, this will delay many vehicles on the other side whose individual waiting time may be less than the waiting time of fewer vehicles.
3. In order to disseminate and distribute congestion, it is important to consider cumulative waiting time for a direction. This approach helps addressing the needs of larger number of commuters.
4. This approach addresses balancing the vehicle flow rate and resulting buildup of vehicular density.
5. All things being equal, priority must be given to direction which can evacuate faster.
6. To avoid any direction with higher vehicular flow monopolizing a signal phase, an upper bound is defined beyond which the signal phase must be withdrawn.

3.4 Cumulative Waiting Time

$$\tau = \sum_{1}^{N} \left(\int_{T1}^{T2} dt \right)$$

τ = Cumulative waiting time (CWT) of vehicles for a direction; $T1$ = Entry time of vehicle; $T2$ = Time at computation; N = Number of vehicles.

The other strategy is to incorporate CWT of total area of waiting vehicles, CWTA of vehicle areas, which is road occupancy, then:

Table 1 Priority table

ARCA	CWT	VC	RO	Priority
0	0	0	0	1
0	0	0	1	2
0	0	1	0	3
0	0	1	1	4
0	1	0	0	5
0	1	0	1	6
0	1	1	0	7
0	1	1	1	8
1	0	0	0	9
1	0	0	1	10
1	0	1	0	11
1	0	1	1	12
1	1	0	0	13
1	1	0	1	14
1	1	1	0	15
1	1	1	1	16

ARCA Available road capacity ahead; *RO* road occupancy; *VC* vehicle count and priority of resolution, introduces and explains these terms exhaustively [32]

$$\tau = \sum_{1}^{N} A * \left(\int_{T1}^{T2} dt \right)$$

A = Area of vehicle under consideration.

Road segment static capacity is pre-computed and available from infrastructure database. Every vehicle periodically updates its location through lat. / long., obtained by the GPS application. Dynamic road capacity is continuously computed from incident vehicles on the road segment. Congestion condition is detected by a combination of number of vehicles on a road segment, road occupancy, and average speed of vehicles on the road segment. Priority matrix with highest priority being 16 is defined in Table 1.

4 Algorithm

4.1 Data Structure

City has number of traffic junctions. A traffic junction has number of poles, decided by number of converging roads at a junction; realistically two to eight. Each pole has signals to indicate three movements viz. straight, left, and right; each direction has red, orange, and green indications. Above all, there is a traffic indicator for pedestrian too. Every traffic pole is related to a road segment. Road segment characterizes its capacity or area and quality by achievable speed on the road segment. Road segment identifies its capacity in terms of area which is expressed as length and breadth. For vehicular traffic, breadth is expressed as number of lanes. Road occupancy is expressed by number of vehicles on the road segment and footprint of these vehicles. The functions are expressed as (1) city (traffic junctions), (2) traffic junction (traffic poles), (3) traffic pole (road segment), and (4) road segment (capacity, occupancy, quality) or Road segment (start lat., start long., end lat., end long., linear distance, number of lanes, achievable speed).

Inputs to the algorithm are road occupancy of every road segment. Output is signal phase (junction number, traffic pole, road segment, signal phase duration).

4.2 Algorithm

```
Const
Time to move := t1;
Time to cross the junction := t2;
Main
    For every Road Segment
        Compute RO, VC;
        Compute CWT; ARCA;
    End;
    Sort road segments as per PRIORITY table;
    Junction Consideration;
End
RO, VC
    For all outgoing vehicle
        VC := VC -1;
        RO := RO - Area of Outgoing vehicle;
    End;
    For all incoming vehicle;
        VC := VC +1;
        RO := RO + Area of Incoming vehicle;
    End;
End;
CWT
For every unit time
    CWT := VC + VC;
End;
```

```
ARCA
   ARCA := RC - RO;
End;

Junction consideration
For all Junction directions
   Sort by CWT;
   Sort by Max Mod(RO-ARCA);
   CSP fir Dir 1;
   If waiting vehicles on Dir 2 ≠ 0 then CSP
   Else
   If waiting vehicle on alternate 2 ≠ 0then CSP
   Else
   Hand it over to next direction
   End
   Compute signal phases CSP
End;

CSP
   VC := f(RO);
   Green Phase Dir 1 := t1 + t2* VC;
End;
```

5 Future Work

The signaling information must be pushed to the in-vehicle display to offer preemptive signaling information. This will alert the driver about signal phases and who can then control the mobility of the vehicle. This feature will pave the way for removal of traffic islands, thereby making space and lending to the already crammed-up infrastructure. Secondly, it will drastically reduce capital costs of signaling electronics and operating costs for its maintenance and upkeep. Besides, accurate signaling system will detect traffic offenses, thereby punishing the offenders without any manpower deployment. This will over a period of time eliminate traffic offenses to near nil.

References

1. Kittelson & Associates, Inc. (2008) Texas Transportation Institute, University of Maryland, Siemens ITS, Purdue University, & the Institute of Transportation Engineers, June 2008, Signal timing manual
2. Irwin WC (1977) Planning and construction of a computerized traffic control system. Chattanooga, Tennessee

3. Xiaohong L, Shuhua R, Lan S (2008) A new traffic signal coordinated control method based on mobile agent technology. In: Proceedings of the 27th chinese control conference, Kunming, Yunnan, China, Department of Computer Science and Engineering, Dalian Polytech University, Dalian 116034, P. R. China
4. Zhang W, Tan G, Ding N, Wang G (2012) Traffic congestion evaluation and signal control optimization based on wireless sensor networks: model and algorithms (School of Computer Science and Technology, Dalian University of Technology, Dalian, China
5. Chapter 15 High-order models. http://people.umass.edu/ndh/TFT/Ch15%20High.pdf
6. List GF, Cetin M (2004) Modeling traffic signal control using Petri Nets. IEEE Trans Intell Transp Syst 5(3)
7. dos Santos Soares M, Vrancken J (2008) Responsive traffic signals designed with Petri Nets. In: IEEE international conference on systems, man and cybernetics (SMC)
8. Chin YK, Kow WY, Khong WL, Tan MK, Teo KTK (2012) Q-learning traffic signal optimization within multiple intersections traffic network. In: UK Sim-AMSS 6th european modeling symposium, modeling, simulation & computing laboratory, material & mineral research unit. School of Engineering and Information Technology, University Malaysia Sabah, Kota Kinabalu, Malaysia
9. Anokye M, Abdul-Aziz AR, Annin K, Oduro FT (2013) Application of queuing theory to vehicular traffic at signalized intersection in Kumasi-Ashanti Region, Ghana. Am Int J Contemp Res 3(7)
10. Komada K, Nagatani T (2009) Modeling and simulation for vehicular traffic in city network controlled by signals. TIC-STH, Department of Mechanical Engineering, Shizuoka University, Hamamatsu, Japan
11. Parmar RS, Trivedi B (2017) Modulating traffic signal phases to realize real-time traffic control system. J Transp Technol 7(1):26–35
12. Zhou B (2013) A real-time traffic-responsive strategy for road congestion problem. In: 2013 international conference on computational and information sciences. IEEE. College of Information Science and Technology, Zhejiang Shuren University, Hangzhou, China. https://doi.org/10.1109/iccis.2013.304
13. Wenjie C, Lifeng C, Zhanglong C, Shiliang TU (2005) A real time dynamic traffic control system based on wireless sensor network. In: Proceedings of the 2005 international conference on parallel processing workshops (ICPPW'05). IEEE
14. Lan C-L, Chang G-L (2015) A traffic signal optimization model for intersections experiencing heavy scooter–vehicle mixed traffic flows. IEEE Trans Intell Transp Syst 16(4)
15. Zhang W, Tan G, Ding N, Wang G (2012) Traffic congestion evaluation and signal control optimization based on wireless sensor networks: model and algorithms. In: Mathematical problems in engineering. Hindawi Publishing Corporation, vol 2012, Article ID 573171, 17p. School of Computer Science and Technology, Dalian University of Technology, Dalian, China. https://doi.org/10.1155/2012/573171
16. Chang J-J, Li Y-H, Liao W, Chang, I-C (2012) Intersection based routing for urban vehicular communication with traffic light consideration. IEEE Wirel Commun
17. Yinfei L (2009) Research on synchronizing traffic signals for an Urban arterial road. In: Third international symposium on intelligent information technology application. School of Statistics and Mathematics, Zhejiang Gongshang University, Hangzhou, China
18. Dotoli M, Fanti MP, Meloni C (2004) Coordination and real time optimization of signal timing plans for urban traffic control. In: Proceedings of the 2004 IEEE international conference on networking, sensing 8 control, 21–23 Mar 2004
19. Borkar P, Malik LG (2013) Speed range prediction for subsequent intersections. In: Proceedings of 7th international conference on intelligent systems and control (ISCO)

20. Hu Z, Motani M (2012) DVS: a distributed virtual signboard for information dissemination and preservation in vehicular networks. In: 15th international IEEE conference on intelligent transportation systems, Anchorage, Alaska, USA, 16–19 Sept 2012
21. Dong L, Chen W (2010) Real-time traffic signal timing for urban road multi-intersection. Intell Inf Manag 2:483–486. College of Mechanical Engineering, Shanghai University of Engineering Science, Shanghai, China. http://www.SciRP.org/journal/iim
22. Ran Q, Yang J (2012) A novel closed-loop feedback traffic signal control strategy at an isolated intersection. In: 2012 IEEE international conference on information science and technology, Wuhan, Hubei, China, 23–25 Mar 2012
23. Kponyo JJ, Kuang Y, Li Z (2012) Real time status collection and dynamic vehicular traffic control using ant colony optimization. In: 2012 international conference on computational problem-solving (ICCP), Oct 2012
24. Shladover SE, Li J-Q (2011) Evaluation of probe vehicle sampling strategies for traffic signal control. In: 2011 14th international IEEE conference on intelligent transportation systems, Washington, DC, USA, 5–7 Oct 2011
25. https://en.wikipedia.org/wiki/Google_Traffic
26. http://www.scoot-utc.com/
27. http://www.scats.com.au/
28. Shelby SG, Bullock DM, Gettman D, Ghaman RS, Sabra ZA, Soyke N (2008) An overview and performance evaluation of ACS lite—a low cost adaptive signal control system. Submitted to the 87th TRB annual meeting in Washington, DC, Jan 2008
29. RHODES—real time hierarchical optimized distributed effective system (2008). http://ocw.nctu.edu.tw/course/sc011/2012-08-23-1.pdf
30. Parmar RS, Trivedi B (2014) Identification of parameters and sensor technology for vehicular traffic—a survey. IJTTE Int J Traffic Transp Eng 3(2):101–106
31. Parmar RS, Trivedi B (2014) Real time computation of optimal signal timing to maximize vehicular throughput for a traffic junction. In: 3rd international conference on eco-friendly computing and communication systems (ICECCS 2014). NITK Surathkal, Mangalore, India, pp 194–199, 8–21 Dec 2014
32. Parmar RS, Trivedi B (2017) Shortest alternate path discovery through recursive bounding box pruning. J Transp Technol 1–14

Rajendra S Parmar (Raj) was born in Ahmedabad, India, in 1956. He received his B.Tech. from IIT (BHU), Varanasi, India, in 1978 and M.Tech. in Computer Science, IIT Bombay, Maharashtra, India, in 1992. He started his career with Bhabha Atomic Research Centre (BARC) as Scientific Officer C. Subsequently, he forayed in print media and contributed with pioneering work in the field of Indian language text editors, multilocation simultaneous printing before occupying the position of General Manager at Bennett Coleman & Company Limited (BCCL), one of the top ten newspaper publishers in the world. Later, he moved to BPL Telecommunication, leveraging the company to achieve leadership position in India. He founded Recreate Solutions which managed rendering Rich Media, a form derived by creating format agnostic content. His research interest includes traffic decongestion and IoT.

Prof. Bhushan Trivedi, Ph.D. is working as a Director, GLS Institute of Computer Technology (GLSICT), and the Dean, School of Computer Technology, GLS University. He is also dean zone-1 in Gujarat Technological University, Gujarat, for MCA. He obtained his MCA from MS University, Vadodara, in 1984. He has completed his Ph.D. from Hemchandracharya North Gujarat University in 2008. Three of his books are published by Oxford University Press. He has been part of many technical program committees of reputed journals and conferences. He has published about 87 research papers in various national and international journals and conferences. His research interest includes pedagogy, security, intrusion detection and prevention, expert systems, and neural networks. He is an active Life Member of Computer Society of India and was Chairman of Ahmedabad Chapter in 2007. He is part of many program and organizing committees of CSI Ahmedabad Chapter. He is also a Senior Member of Academy of Computing Machinery (ACM). He received an award for the work on effective teaching by IUCEE in 2009. He is also given Chapter Patron Award by Computer Society of India in 2011.is working as a director at GLS Institute of Computer Technology (GLSICT) and the dean, school of computer technology at GLS University. He is also dean zone-1 in Gujarat Technological University, Gujarat for MCA. He is an MCA from MS University, Vadodara from 1984. He has completed his Ph D from Hemchandracharya North Gujarat University in 2008. Three of his books are published by Oxford University Press. Prof. Trivedi has been part of many Technical Program Committees of reputed journals and conferences. Prof. Trivedi has published about 87 research papers in various national and International journals and conferences. His research interest includes pedagogy, security, intrusion detection and prevention, expert systems and neural networks. Prof Trivedi is an active life member of Computer Society of India, was chairman Ahmedabad Chapter in 2007. He is part of many program and organizing committees of CSI Ahmedabad Chapter. He is also a senior member of Academy of Computing Machinery (ACM). Prof. Trivedi received an award for the work on effective teaching by IUCEE in 2009. He is also given Chapter Petron award by Computer Society of India in 2011.

Discrete Wavelet Transform and kNN-Based Fault Detector and Classifier for PV Integrated Microgrid

Murli Manohar, Ebha Koley, Yuvraj Kumar and Subhojit Ghosh

Abstract The growing penetration of distributed energy resources (DERs) in modern power distribution networks operating as microgrid poses a great challenge for the conventional protection scheme due to significant variation in the fault current levels under the grid-connected and islanded mode of operation. In this regard, this paper has devised an efficient protection scheme based on discrete wavelet transform (DWT) and k-nearest neighbour (kNN) for fault detection/classification implemented for dual modes of microgrid operation considering the photovoltaic PV source and nonlinearity in the load. The proposed approach utilizes the three-phase voltage and current signals obtained during shunt faults in the distribution line under widely varying fault parameters. The pre-processing of signals through DWT determines the approximate coefficient. The standard deviation (SD) of the approximate coefficient so obtained is further fed as the input to the kNN-based classifier for fault detection/classification task separately for grid-connected and islanded mode. The test result analysis clearly reveals the effectiveness of the proposed approach and hence validates the performance.

Keywords Microgrid · Distributed energy resources (DERs) · Discrete wavelet transform (DWT) · k-nearest neighbour (kNN) · Photovoltaic (PV) · Fault detection and classification

M. Manohar (✉) · E. Koley · Y. Kumar · S. Ghosh
National Institute of Technology, Raipur, India
e-mail: murlimanohar2311@gmail.com

E. Koley
e-mail: ekoley.ele@nitrr.ac.in

Y. Kumar
e-mail: yuvrajkumarsinha@gmail.com

S. Ghosh
e-mail: sghosh.ele@nitrr.ac.in

© Springer Nature Singapore Pte Ltd. 2018
M. L. Kolhe et al. (eds.), *Advances in Data and Information Sciences*, Lecture Notes in Networks and Systems 38, https://doi.org/10.1007/978-981-10-8360-0_2

1 Introduction

With the advent of renewable energy resources such as photovoltaic (PV), wind energy, modern power distribution networks have witnessed a gradual transition of central generation towards the distributed one. The concept of introducing distributed generation has led to the significant reduction in power transmission and distribution burden, thereby increasing the reliability and overall efficiency of the system. The penetration of DERs in the distribution network to cater to the power demand at consumer end has contributed greatly to establish a single controllable entity named as microgrid [1, 2]. The power flow in microgrid is bidirectional which enable it to operate under dual mode, i.e. grid-connected and islanding mode. The islanding mode refers to the isolated state of microgrid feeding the critical loads under the occurrence of fault in the grid.

The fault current variation under dual mode operation of microgrid poses great protection challenges. Apart from this, the fault current carrying capability of inverter-based DERs is only 2–3 times of the rated current, whereas it is up to 10 times for the synchronous-based DERs, which further complicate the protection task up to great extent [3, 4]. The conventionally designed protection schemes based on fundamental component of voltage and current may not be able to identify the type of fault occurred in the system. This has led the motivation to propose such protection technique which can provide fast, reliable and accurate fault detection/classification task under fault in the microgrid under dual mode of operation.

Several microgrid protection techniques have been mentioned in the literatures which have been declared as the possible solution to the persisting protection issues. Some of them include differential energy-based microgrid protection [4], comprehensive protection strategy for islanded microgrid [5], combined S-transform and data-mining-based protection scheme [6], a differential zone protection scheme [7], travelling wave-based protection scheme [8] and time-frequency transform-based differential scheme [9]. The protection challenge addressed in [10] has mainly focused on high impedance fault detection, whereas in [6], two different decision tree-based classifiers have been used for fault detection and classification. Neither [7] nor [8] have focussed on fault classification. In [11] also, the technique has been reported on detection only and that too without taking into consideration the operating modes of microgrid. Effective feature selection is very important criteria during the development of a data-mining technique. In this regard, the discrete wavelet transform (DWT) used as feature extraction technique in the proposed study carries the advantage over above-reported techniques in analyzing the non-stationary (transient) signals. Despite the significant contribution of the above-mentioned proposed techniques in the field of microgrid protection, the intermittent behaviour of renewable DERs and nonlinearity in the load is still left to be addressed in the actual fault scenario. The actual distribution network comprises large interconnection of nonlinear nature of loads in the system. Also, the effect of power quality disturbances associated with inverter-interfaced DERs and nonlinear characteristics of load cannot be ignored. Keeping these in mind, this paper has got the motivation to propose an accurate and reliable fault detection/classification technique based on discrete wavelet transform (DWT) and k-nearest neighbour (kNN) approach. The proposed technique utilizes standard deviation of the approximate coefficients extracted by pre-processing of voltage and current signals collected at relaying bus through DWT, as the input feature to train the kNN-based classifier. The PV intermittency and nonlinearity in the load have been considered while creating the fault scenarios. kNN is a very simple and widely known algorithm whose classification strategy is based on statistical approach. Unlike the other supervised learning algorithms which require explicit training of data, kNN carries the advantage of making use of the local contexts retrieved from training sample to declare the classification result [12, 13]. The performance of proposed technique has been examined and validated against various test cases generated with wide variations in fault parameters under both modes of operation and the summary of results have been presented.

The remaining paper is organized into following sections: The modelling and simulation of microgrid system under study in the MATLAB/Simulink environment are illustrated in Sect. 2. Section 3 describes the feature extraction technique using DWT and the development of proposed kNN-based fault detector/classifier. Test results are summarized and analyzed in Sect. 4 followed by conclusion in Sect. 5.

2 Modelling and Simulation of Microgrid Under Study

Figure 1 represents the single line layout of the 34.5 kV, 60 Hz microgrid under study simulated using MATLAB/Simulink. This microgrid incorporates the combination of two DERs namely synchronous-based source SDG and photovoltaic energy resource, PVDG connected to the network at bus B3 and B4, respectively [14]. S1, S2, S3 and S4 represent the four sections of the line with total line length of 20 km. P1, P2, P3, P4 and P5 represent the combination of linear and nonlinear loads. Bus B1 acts as the relaying bus at which the raw voltage and current signals are retrieved for processing. Voltage source converters (VSCs) with converter control unit, maximum power point tracking (MPPT) and phase-locked loop (PLL) and transformer have been used to interface PVDG with the AC grid. During grid-connected operation, SDG remains disconnected from the network, whereas in islanded mode, both DGs feed the loads. The variation in BG fault current magnitude at bus B1 for grid-connected and islanded mode is depicted in Fig. 2.

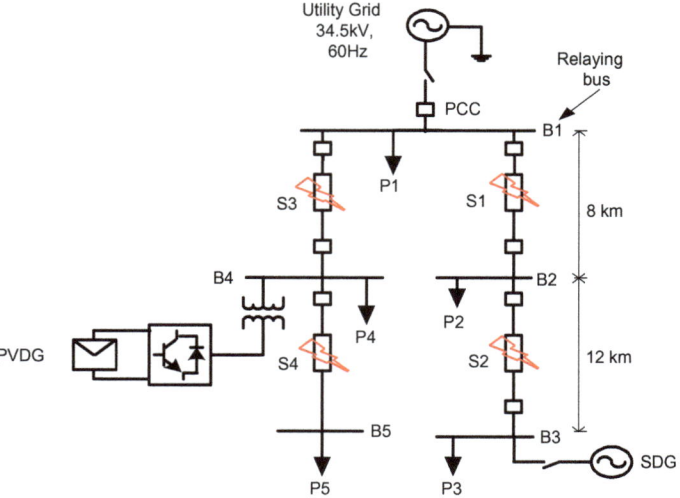

Fig. 1 Single line layout of microgrid model under study

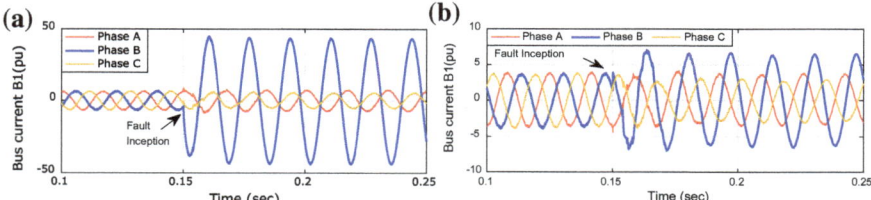

Fig. 2 Current at bus B1 due to BG fault inception in section S3 with fault resistance $R_f = 5\ \Omega$ and location $L_f = 3$ km under, **a** grid-connected mode, **b** islanded mode

3 Development of kNN-Based Fault Detector/Classifier

3.1 Overview of the Presented Technique

The objective of imparting protection to the microgrid is to ensure accuracy and reliability in the detection and classification of fault for immediate isolation of faulty feeder. Keeping this in mind, a discrete wavelet transform and kNN-based fault detector/classifier have been devised and presented in the paper which can perform efficiently under intermittent nature of PVDG and nonlinearity in the load as well. The different steps of the proposed scheme (Fig. 3) can be sketched as follows:

Step 1: Voltage and current signals are retrieved at bus B1 by creating the fault scenarios with variation in fault parameters such as fault resistance (0–100 Ω), inception angle (0° to 90°) and fault location (0–20 km).

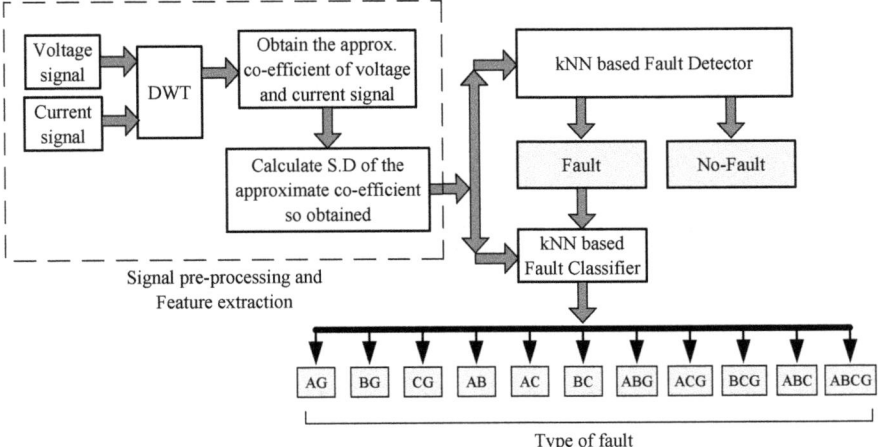

Fig. 3 Schematic view of kNN-based fault detector/classifier

Step 2: The signals obtained in step 1 is pre-processed through low-pass Butterworth filter at the cut-off frequency of 480 Hz.

Step 3: The processed signal is sampled with sampling frequency of 1.2 kHz.

Step 4: The approximate coefficients are extracted through discrete wavelet transform (DWT) using db3 wavelets up to third level of decomposition and standard deviation is obtained which is further fed as the input feature for kNN.

Step 5: Select the number of nearest neighbours and train the dataset using kNN.

Step 6: Any fault sample is tested through the trained network and depending on the output of respective kNN, the relay trip signal is generated to the respective breaker for isolation of faulty phase in the network.

3.2 Pre-processing of Signal and Feature Extraction

Wavelet transform is very dominant and versatile tool to analyze and record the pattern obtained due to any disturbance or fault occurred in the system. The advantage of wavelet transform over Fourier transform lies in the fact that latter can be utilized to analyze the signal only in the frequency domain. So, wavelet transform is considered better than Fourier transform. Wavelet transform split the signal into different scales corresponding to distinct frequency patterns through translation (shift in time) and dilation (compression in time) of mother wavelet [6, 11]. The mother wavelet can be described by following equation:

$$\emptyset_{jk}(t) = 2^{-j/2} \phi \left(2^{-j} t - k \right) \tag{1}$$

Signal is analyzed by wavelets through multi-resolution technique at distinct resolution and frequencies. The decomposition of input signal $x(n)$ through low-pass Butterworth filter is carried out by the following equation:

$$y_{low}[k] = \sum x[n].h[2k - n] \tag{2}$$

where $Y_{low}[k]$ is the output of low-pass filter $h[n]$ obtained after down sampling by 2. A total number of 6 features including B1 bus voltage and current in section S1 for all the three phases A, B and C have been considered for feature extraction process.

3.3 Fault Detection/Classification Using kNN

Pre-processing and feature extraction are followed by the development of kNN-based technique for performing the required task of fault detection/classification separately under both modes of microgrid operation.

kNN algorithm is very simple and widely known among all the other pattern recognition techniques due to its flexible decision boundaries. It exhibits lazy learning approach; i.e. it imparts negligibly less effort at the time of training and exceptionally full effort during the prediction [12, 13]. The k-nearest neighbour performs the classification task on the basis of similarity index by taking into account the distance measure in which 'k' refers to the integer value mostly lying in the range of 3–10. It is preferred to choose odd value of k to avoid the situations of tie during the prediction of any class. Any particular class among all available input classes is predicted as the output by kNN on the basis of majority votes casted by the neighbouring points corresponding to the nearer class. For any test case X, the probability of the case belonging to the class C_i should be maximum as given by the equation:

$$kNN(X) = \max p\,(C_i, X) \tag{3}$$

For performing the task of classification, weights are given to the neighbours in such a way that the nearer neighbours add up to more weights to the average than that of the farther one [15]. Weights are assigned to the nearest neighbours on the basis of Euclidean distance calculated using the following equation:

$$d\,(x, t) = \sqrt{(x_1 - t_1)^2 + (x_2 - t_2)^2 + \cdots + (x_p - t_p)^2} = \sqrt{\sum_{i=1}^{p} (x_i - t_i)^2} \tag{4}$$

The kNN algorithm-based protection technique performs the task of fault detection and classification under grid-connected and islanded mode separately. The fault detector has two targets namely fault and no-fault, whereas the fault classifier has 11 assigned target outputs namely AG, BG, CG, AB, AC, BC, ABG, ACG, BCG, ABC and ABCG.

4 Performance Evaluation of Proposed Scheme

Training of the dataset consisting of features obtained through DWT is followed by testing of developed kNN-based fault detector/classifier by randomly generated testing cases considering the variation in fault parameters. The patterns of parameters for testing samples were different from that considered during training. The detail of training and testing samples generated are summarized in Table 1. Total 1388 samples for training and 754 samples for testing purpose were generated under both operating mode of microgrid. The dataset contains no-fault cases of 100 and 25 simulated, respectively, for training and testing.

The appropriateness of developed protection technique for microgrid has been examined through testing under both modes of operation separately. The test results have been depicted in Table 2. For each type of fault, 64 fault and 50 no-fault testing cases were generated collectively for both modes and tested with kNN classifier. The accuracy of fault detector/classifier (FD/C) in each case was determined to be 100% as can be observed in Table 2. This clearly demonstrates the accuracy and reliability of the proposed scheme in providing protection of microgrid against shunt faults under both modes of operation.

5 Conclusion

The protection technique devised in the paper is based on wavelet transform and kNN-based fault detection/classification approach which is able to operate under both modes of microgrid. The input features containing the standard deviation of approximate coefficients of voltage and current signal at the relaying bus are obtained through pre-processing of the signals by DWT. The features so obtained are fed as input to train the kNN-based classifier to accomplish the task of fault detection/classification successfully. The proposed scheme has not been influenced by nonlinearity in the load and intermittency of the PV source. Based on the testing performance, the proposed protection technique has been found to be efficient, accurate and reliable against shunt fault detection/classification with accuracy of 100% under both modes of operation. Due to the successful implementation of the proposed protection technique in detection/classification of faults, it can also be extended further in future to carry out other protection tasks related to microgrid such as faulty zone identification and location estimation.

Table 1 Detail of training and testing data generation

Fault parameters	Training data generation				Testing data generation			
	Specification	Mode of operation		Total training samples	Specification	Mode of operation		Total testing samples
		Grid-connected	Islanding			Grid-connected	Islanding	
Fault types	LG, LL, LLG, LLL, LLLG	11	11	2 * [(11 * 3 * 2 * 9) + 100] = 1388 samples	LG, LL, LLG, LLL, LLLG	11	11	2 * [(11 * 2 * 2 * 8) + 25] = 754 samples
Fault resistance	0, 50, 100 Ω	3	3		25, 75 Ω	2	2	
Inception angle	0°, 90°	2	2		30°, 60°	2	2	
Fault location	2, 4, 6, 8, 10, 12, 14, 16, 18 km	9	9		3, 5, 7, 9, 11, 13, 15, 17 km	8	8	
No-fault cases	–	100	100		–	25	25	

Table 2 Performance of protection scheme against different type of fault with variation in fault parameters

Fault type	Output of fault detector/classifier (FD/C)										Total no. of test cases	Total no. of correctly predicted cases	Accuracy (%)
	Grid-connected mode					Islanding mode							
	LG	LL	LLG	LLL	LLLG	LG	LL	LLG	LLL	LLLG			
AG	AG	NF	NF	NF	NF	AG	NF	NF	NF	NF	64	64	100
BG	BG	NF	NF	NF	NF	BG	NF	NF	NF	NF	64	64	100
CG	CG	NF	NF	NF	NF	CG	NF	NF	NF	NF	64	64	100
AB	NF	AB	NF	NF	NF	NF	AB	NF	NF	NF	64	64	100
AC	NF	AC	NF	NF	NF	NF	AC	NF	NF	NF	64	64	100
BC	NF	BC	NF	NF	NF	NF	BC	NF	NF	NF	64	64	100
ABG	NF	NF	ABG	NF	NF	NF	NF	ABG	NF	NF	64	64	100
ACG	NF	NF	ACG	NF	NF	NF	NF	ACG	NF	NF	64	64	100
BCG	NF	NF	BCG	NF	NF	NF	NF	BCG	NF	NF	64	64	100
ABC	NF	NF	NF	ABC	NF	NF	NF	NF	ABC	NF	64	64	100
ABCG	NF	NF	NF	NF	ABCG	NF	NF	NF	NF	ABCG	64	64	100
NF	NF	NF	NF	NF	NF	NF	NF	NF	NF	NF	50	50	100

References

1. Hosseini SA, Abyaneh HA, Sadeghi SHH, Razavi F, Nasiri A (2016) An overview of microgrid protection methods and the factors involved. Renew Sustain Energy Rev 64:174–186
2. Lasseter RH (2002) MicroGrids. In: IEEE Conference, 2002, pp 305–308
3. Basak P, Chowdhury S, Halder Nee Dey S, Chowdhury SP (2012) A literature review on integration of distributed energy resources in the perspective of control, protection and stability of microgrid. Renew Sustain Energy Rev 16(8):5545–5556
4. Kar S, Samantaray SR, Zadeh MD (2015) Data-mining model based intelligent differential microgrid protection scheme. IEEE Syst J PP(99):1–9
5. Lai K, Haj-ahmed MA (2016) Comprehensive protection strategy for an islanded microgrid using intelligent relays. IEEE Trans Ind Appl
6. Kar S, Samantaray SR (2014) Combined S-transform and data-mining based intelligent microgrid protection scheme. In: IEEE Conference, 2014, pp 1–5
7. Sortomme E, Ren J, Venkata SS (2013) A differential zone protection scheme for microgrids. In: IEEE Conference, 2013, pp 1–5
8. Li X, Dysko A, Burt GM (2014) Traveling wave-based protection scheme for inverter-dominated microgrid using mathematical morphology. IEEE Trans Smart Grid 5(5):2211–2218
9. Kar S, Samantaray SR (2014) Time-frequency transform-based differential scheme for microgrid protection. IET Gener Transm Distrib 8(2):310–320
10. Samantaray SR, Joos G, Kamwa I (2011) Differential energy based microgrid protection against fault conditions. In: IEEE Conference, 2011, pp 1–7
11. Lin H (2015) Distance protection for microgrids in distribution system. In: IEEE Conference, 2015, pp 731–736
12. Yu X, Yu X (2007) Novel text classification based on k-nearest neighbor, pp 19–22
13. Yu Z, Chen H, Liu J, You J, Leung H, Han G (2016) Hybrid k-nearest neighbor classifier. IEEE Trans Cybern 46(6):1263–1275
14. Hooshyar A, El-Saadany EF (2016) Fault type classification in microgrids including photovoltaic DGs. IEEE Trans Smart Grid 7(5):2218–2229
15. Budnik M, Pozniak-koszalka I, Koszalka L (2012) The usage of the k-nearest neighbour classifier with classifier ensemble. In 2012 12th international conference on computational science and its applications (ICCSA), no 1, pp 170–173

Automated Tool for Extraction of Software Fault Data

Pradeep Singh and Shrish Verma

Abstract Open-source software repositories contain lots of useful information related to software development, software design, and software's common error patterns. To access the software quality an automated software fault data extraction and preparation, which can be used for further prediction is still a major issue. Prediction of software fault has recently attracted the attention of software engineers. These prediction models require training fault data of projects. The fault training data contains information of software metrics and related bug information, and these data have to be prepared for each project. But it is not so easy to collect and prepare the fault data for the prediction model. We developed an automatic tool which extracts and prepares fault data for the prediction models. By using these automatic tools, we have extracted the data from the open-source projects developed in various languages. Extraction of fault data of various projects which includes source code and related defects from open-source software repository is performed. Various versions of open-source project software were taken from source forge and used for this purpose.

Keywords Software metrics · Defects · Open-source software

1 Introduction

Software repositories are the collection of artifacts in the process of evolution of software systems. Software repositories often contain data over many years of development of a software project. These data typically are versions of the project, who made changes, why, and what changes were made, and when those changes were

P. Singh (✉)
Department of Computer Science and Engineering, National Institute of Technology,
Raipur, India
e-mail: psingh.cs@nitrr.ac.in

S. Verma
Electronics & Telecommunication Engineering, National Institute of Technology, Raipur, India
e-mail: shrishverma@nitrr.ac.in

© Springer Nature Singapore Pte Ltd. 2018   29
M. L. Kolhe et al. (eds.), *Advances in Data and Information Sciences*, Lecture Notes
in Networks and Systems 38, https://doi.org/10.1007/978-981-10-8360-0_3

made, etc. The repository usually stores information in the form of a file system tree for the system which is hierarchy of files and directories. Any number of clients can connect to the repository, and then read or write to these files. Some common types of software repositories are source code version control system repositories, bug repositories, and communication archives, etc.

For software quality assessment, software fault data are needed as input to the models. Software fault data consists of defect per module and related software measurements. So an automatic tool for extraction of fault data and software measurement data is required. A lot of testing tools are available freely on the Internet for researchers to calculate metric data, but calculating fault data is tedious task due to lack of appropriate tools for fault data collection and most of defect data needs to be calculated manually. In this paper, we describe automatic tool development for extraction of the relevant fault data of various open-source projects from www.sourceforge.net.

2 Related Work

A lot of research has been carried out in the area of change data collection and defect data collection. Chidamber and Kemerer initially proposed object-oriented metrics, which measure software quality attributes [1]. Watanabe et al. have extracted the software fault data of Sakura Editor manually [2]. Software fault data helps in software quality maintenance and also helps in timely delivery of the software by using software defect prediction. Mahajan et al. [3] have used Bayesian regularization (BR) technique for finding the software faults before the testing process. This technique helps to reduce the cost of software testing which reduces the cost of the software project. The basic purpose of BR technique is to minimize a combination of squared errors and weights, and then determine the correct combination so as to produce an efficient network. BR technique algorithm-based neural network tool is used for finding the results on the given public dataset. Singh et al. [4] developed a framework for automatic extraction of human understandable fuzzy rules for software fault detection/classification. This is an integrated framework to simultaneously identify useful determinants (attributes) of faults and fuzzy rules using those attributes. Bishnu and Bhattacherjee [5] have evaluated quadtree-based k-means clustering algorithm for predicting faulty software modules. Kaur et al. [6] have proposed k-Sorensen means algorithm, they have used Sorensen measure to calculate the distance to form a cluster. They used the NASA MDP datasets, trained and tested using JM1, PC1, and CM1, and three metrics namely requirement metrics, static code metrics and alliance metrics have been used and thus concluded that alliance metrics are best for prediction model. An approach for software defect data extraction from open-source repository (OSR) is proposed by [7], in which the defect information extracted from Github. DuoTracker, a tool to track and analyze software defects for software process assessment for the quality of individual software engineer's work is proposed in [8].

Duo tracker extracts only defects and still for the various work fault data is collected manually [2]. So there is a need of tool which can efficiently collect the software fault data. A configuration management system (CMS) tool is developed to calculate the changes made during development using log files [9]. But this tool is not capable to map the software defects with its module's software metric. CMS tool can only assist software researchers in collecting defect and change data for software systems. It is not creating the software fault data automatically which is ready to be used for research in software defect prediction. Extracting fault data from change repositories is still a challenging task [10]. A comparative study of three approaches to extract fault data from the changed repository of the barcode open-source system has been performed by [10], and it was found that the confidence in manual system is more than the two automatic approaches. Hall et al. concluded that it is still very difficult to reliably extract fault fixing data from change repositories, using automatic tools [10].

3 Subversion

Subversion is a free/open-source version control system. SVN provides facility of multiple streams to make modifications simultaneously in order to progress quickly. Due to the version management, there is no need to fear about losing some useful feature. If some incorrect change is made to the data, it can be undone. A group of changes at the same time is called a revision which creates a new state of the file system tree. The details of each revision are recorded in the shared changelog with author, date, list of files affected/changed, and log message indicating the type of revision, etc. Each revision is assigned a unique natural number, one greater than the number assigned to the previous revision. The evolution data is included in the changelog of each project containing years of fault fix revisions, and hence they can be used as an indicator for software faults. In this study, Sourceforge.net is used for project data extraction which is a Web-based source code repository.

4 Fault Data Extraction Process

In order to perform extraction, a special piece of software called a Subversion client was installed. Subversion clients are readily available for various operating systems. The TortoiseSVN [11] is used to extract raw data from SCM systems including changelogs, source code, bug info, and number of faulty modules. TortoiseSVN is free software for software developers (programmers) and is available at. TortoiseSVN is a Subversion client, implemented as a Microsoft Windows shell extension. It is released under the General Public License (GNU). The TortoiseSVN is used to access the complete online software repository. Each software project has a "URL pattern," using the specific pattern the projects repository can be accessed at client side. For example, iText software repository has been accessed by the URL pattern https://itext. svn.sourceforge.net/svnroot/itext. To access the software repository if proxy address setting is required, user can specify the proxy address and port number. Once the software repository or selected project is retrieved on the local machine, complete source code and all revision logs become available locally. Out of these locally available data, extraction module extracts source code, various software metrics, and bug (fault) information. There are numerous open-source and commercial tools available on the Web which can generate object-oriented metric, complexity metric, and some other static code metrics like CCCC, SciTools (understand) [12], prest [13], Ckjm [14].

The projects used in the study have multiple releases of the same software system. The data of changes and improvements to existing functionality as well as additions of new features for two or more releases of a software project is available in the repository. In the current work, a novel tool to extract software fault data from various modules of an open-source repository is developed from scratch which is now described hereunder. Our objective is to find the error-prone files based on metric data, so we need metric data and defect data. In order to do this, we have taken the various releases of the same projects.

5   Software Metric Extraction

In this section, we define the metrics that we extracted from the source code of the project. These metrics are classified into three groups: design, code, and others. The metric suite extracted is dependent upon the language used for development of the software. In case of object-oriented language, we have addition object-oriented metrics such as CK metrics in addition to Halstead software science metrics and Loc-based metrics. These metric suites offer informative insight into when developers are following object-oriented principles in their design. CK metrics have generated a significant amount of interest and are adopted by practitioners and are also being incorporated into industrial software development tools such as Rational Rose and Together.

Computation of the number of bugs by using SVN repositories of each project is done by identification of the project with release point. Extraction of the log data from each file in the projects is done by using log subcommand. Finally, searching the word (bugs or fixed) and counting the frequency of appearances is performed. Next, we have to associate the bugs to the classes and files found in the source code. We did this one at a time for each bug in each affected version. We examined the bugs one by one and, in each version concerned, we searched bug between the intervals of two version release date and associated that bug with the source code of the previous version. If we found such a class/file, then we increased the number of bugs in that class/file. If the bug fix changed more than one class, then the bug was associated with all these classes. For the java project, we computed the bug count per file and

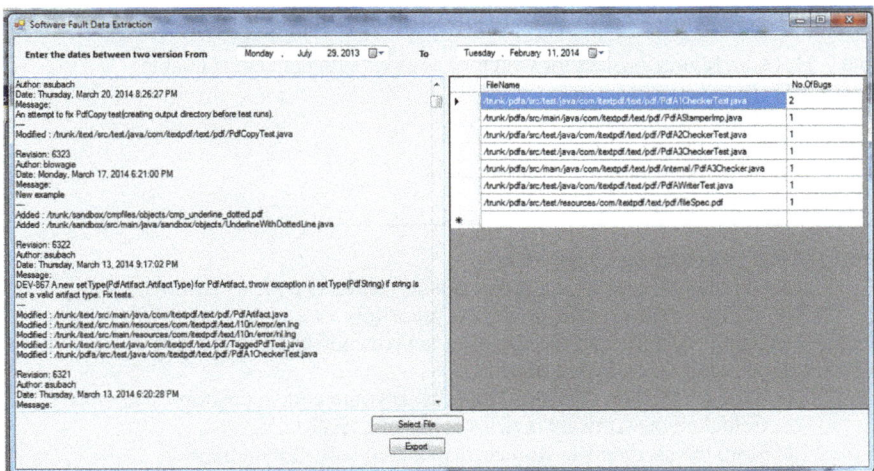

Fig. 1 Interface to specify dates of two versions for which the bugs are to be retrieved

assigned that bug count to the public class. With this method, we extended the tables obtained in the previous section with a new column containing the number of defects for each class.

6 Development of Automated Tool for Extraction of Software Fault Data (ESF Tool)

The metrics data is computed based on the source files of one release and the bug data is computed based on the log data. Our objective is to find the error-prone files based on metric data, so we need metric data and bug data. Source code of both project obtained from sourceforge.net. In order to predict faulty modules using data mining techniques on software repositories, the data pertaining to faults, related fixed modules/files, and software metrics are to be extracted first. So far the process of extraction of software fault data is manual which is tedious and time consuming. In the present work, an automated tool using TortoiseSVN, Visual Basic and MS-Access is developed for the extraction of complete fault data. It is pertinent to mention here that earlier manual retrieval carried out for the project used in [2] takes huge time whereas it takes few seconds only with the help of proposed tool thus giving a huge time saving and accuracy to users. As shown in Fig. 1, the proposed tool automatically extract faulty modules, related software metric, and number of faults reported and fixed between one version of a software to its next version/s. TortoiseSVN is an open-source tool which used to connect online repository to the local one. The proposed tool then extracts the required data from this locally available repository. A pseudo-code elucidating the flow of the proposed tool is given in the Algorithm 1.

Algorithm 1: Software Fault data preparation using Open-source Repository

Input - I1. Open Source project for which various version data are available
        I2. Bug Repository
        I3: Subversion System
Output -O1. Source Code of Project
        O2. Software Metrics of each Module
        O3. Fault Count of each Module

Step 1 (Extract Software repository data)
        1a. Get the online software projects with complete code, revision and logs
        1b. For each online software project repository
        1c. Using SVN, extract the complete source code and change logs
Step 2(Preprocessing for Software Metrics)
        2a. For each version retrieved from the software code repository
        2b. Calculate the software metrics at module level
        2c. Store the module/file with corresponding software metrics
Step 3(Extraction of Software Fault)
        3a. For each version retrieved from the software code repository
        3b. Specify the dates between two revisions to get the change log
        3c. Search the word "BUG/Fix/Fixed" in the log files
        3d. Assign the bug count to the corresponding module/file
Step 4(Mapping the Extraction of Software Fault)
        4a. For each file retrieved from the software code repository
        4b. Match the file name extracted from 2c with the file name in 3c.
        4c. Assign the bug count to all matched files

In order to obtain the required fault data provision for specifying the date is provided in the proposed tool as shown in Fig. 2. Once the dates specified by the user the tool generates the changelog. Changelog is a complex file which includes various text messages. It provides the revision number, author, date, message, and affected files. The message available in the changelogs is important because they contain information of fault and details of all the modules/file affected due to particular fault. The proposed tool parses each message and stores the faulty modules for each reported fault.

An interface is developed to show the revision information, faults, and the files associated with the revision. At the right-hand side of the interface faulty modules with number of bugs extracted after processing is shown in Fig. 1.

Mapping of Software Faulty Modules and Software Metrics:

Once the information about the modules which are faulty is available, the mapping process generates the final file by mapping it with software metrics of each module. For the java project, the bug count per file is computed and that bug count is then assigned to classes. Also, the numbers of bugs fixed are computed by the information presented in the changelog. If any bug is fixed without entering it in the changelog, it is not possible to trace such bugs by the proposed system. The metrics data are computed based on the source files of each release mainly using CKjm and sciTool, and then the bug data are computed based on the log files. The objective here is to

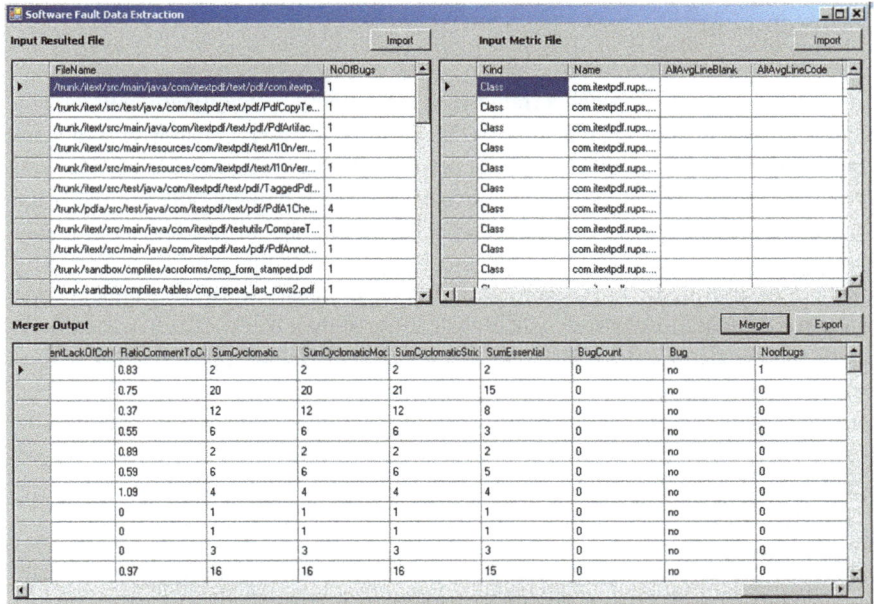

Fig. 2 Interface to mapping the files with metrics and bugs with each module/file

Table 1 Descriptive statistics of few software metrics of FreeOrion

Software metrics	Count line code	No. of functions	Count line blank	No. of pre-processor lines	Executable statements	Ratio comment to code
Min	3	0	4	1	0	0
Max	1924	64	217	27	1069	7.6
Mean	131.36	9.167	34.849	9.063	58.476	0.809
StdDev	217.11	11.992	34.485	5.601	123.024	1.118

find the error-prone modules/files together with their metric data, hence metric data and bug data are mapped with corresponding module. Figure 2 shows the mapping of each software file/module with its metric and faults, which in turn creates complete fault data for analysis.

Software fault data of various SourceForge repositories is extracted. One of the software FreeOrion [15] is analyzed by our ESD system which developed in C++ and python. We have taken two points of release for the project. We extracted the log data from initial point version 0.1 dated 2004-03-07 to Version 3.12 for each file in the projects. Then our tool has extracted metrics of the initial file and extracted bugs from changelog and mapped it to the corresponding software. A brief description of the software metric of the project is shown in Table 1. Total instances or modules are 126 and out of which 15 modules are faulty.

7 Conclusion

In this work, we have described a framework for extraction of software fault data by using open-source software repositories. We have developed an automated tool for extraction of a number of software bugs per module per version and mapping with software metrics of each module/file. The developed tool will not only useful for the research community, but also for the software industry. The tool developed is not only extracting defects but also creating the software fault data that can be used directly for software fault prediction of similar and other projects. Software engineers can also use it for quantitative assessment of the quality of their work. Data extracted with the tool can be used for deciding rules that can be easily incorporated in an IDE as development guidelines. Early software defect prediction models can be developed to predict defect and change prone classes. This will help software practitioners to correctly plan which software module needs more testing resources in early phases of software development.

Acknowledgements This study is partially supported by Chhattisgarh Council of Science and Technology (CGCOST) C.G. under Grant 8068/CCOST. The findings and opinions in this study belong solely to the authors and are not necessarily those of the sponsor.

References

1. Chidamber SR, Kemerer CF (1994) A metrics suite for object oriented design. IEEE Trans Softw Eng 20(6):476–493
2. Watanabe S, Kaiya H, Kaijiri K (2008) Adapting a fault prediction model to allow inter lan-guagereuse. In: Proceedings of 4th international workshop on Predictor models in software engineering—PROMISE'08, p 19
3. Mahajan R, Gupta SK, Bedi RK (2015) Design of software fault prediction model using BR technique. Procedia Comput Sci 46:849–858
4. Singh P, Pal NR, Verma S, Vyas OP (2016) Fuzzy rule-based approach for software fault prediction. IEEE Trans Syst Man Cybern Syst 47(5):1–12
5. Bishnu PS, Bhattacherjee V (2012) Software fault prediction using quad tree-based K-means clustering algorithm. IEEE Trans Knowl Data Eng 24(6):1146–1150
6. Kaur D, Kaur A, Gulati S, Aggarwal M. A clustering algorithm for software fault prediction 1 2
7. Pei H, Ai J (2014) Collecting software defect data automatically from web site of open-source software. In: ICRMS 2014—Proceedings of 2014 10th international conference on reliability, maintainability and safety, pp 333–337
8. Akjnwale O, Dascalu S, Karam M (2006) DuoTracker: tool support for software defect data collection and analysis. CMM
9. Malhotra R, Agrawal A (2014) CMS tool. ACM SIGSOFT Softw Eng Notes 39(1):1–5
10. Hall T, Bowes D, Liebchen G, Wernick P (2010) Evaluating three approaches to extracting fault data from software change repositories. Lecture Notes in Computer Science (including Subser. Lect. Notes Artif. Intell. Lect. Notes Bioinformatics), vol 6156 LNCS, pp 107–115
11. Home · TortoiseSVN (Online) Available: https://tortoisesvn.net/. Accessed 10 Nov 2017
12. SciTools.com (Online) Available: https://scitools.com/. Accessed 10 Nov 2017

13. The community platform for bioinformatics—OMICtools (Online) Available: https://omictools.com/. Accessed 10 Nov 2017
14. Spinellis D (2005) Tool writing: a forgotten art? IEEE Softw 22(4):9–11
15. Freeorion (Online) Available: http://www.freeorion.org/. Accessed 10 Nov 2017

Comparative Study of Mobile Forensic Tools

Animesh Kumar Agrawal, Pallavi Khatri and Sumitra Ranjan Sinha

Abstract Mobile forensics is a field of digital forensics that is galloping at a rapid pace. It encompasses feature phone forensics as well as smartphone forensics. In the early days of mobile forensics, concentration was on somehow extracting data related to feature phones, which included contacts, call logs, SMS that were stored in phone memory and subscriber identity module (SIM) card. Mobile phones are large troves of personal information which if compromised can have a very damaging effect on the individual. As a result, companies are concentrating on securing the data both at rest and in motion. Towards this, encryption technologies are being used to provide robust security to prevent any data sniffing or man-in-the-middle attack. This is making the task of mobile forensics more difficult since companies are providing end-to-end data encryption. As a result, mobile forensics is becoming a nightmare and a big challenge for Law Enforcement Agencies (LEAs). This work presents an experimental study of various mobile data acquisition tools used in past to extract data and proposes a manual method of data extraction that will prove to be an advantage over expensive commercial forensic tools.

Keywords Mobile forensics · Acquisition · Extraction · adb
Commercial tools

1 Introduction

The word "mobile forensics" immediately connects our mind to an Android smartphone, since it is the most popular OS in the mobile fraternity. However, mobile forensics includes basic phones, smartphones and feature phones. And the mobile

A. K. Agrawal · P. Khatri (✉) · S. R. Sinha (✉)
Department of CSE, ITM University, Gwalior, India
e-mail: pallavi.khatri.cse@itmuniversity.ac.in

S. R. Sinha
e-mail: jitsumitra@gmail.com

A. K. Agrawal
e-mail: akag9906@gmail.com

© Springer Nature Singapore Pte Ltd. 2018   39
M. L. Kolhe et al. (eds.), *Advances in Data and Information Sciences*, Lecture Notes
in Networks and Systems 38, https://doi.org/10.1007/978-981-10-8360-0_4

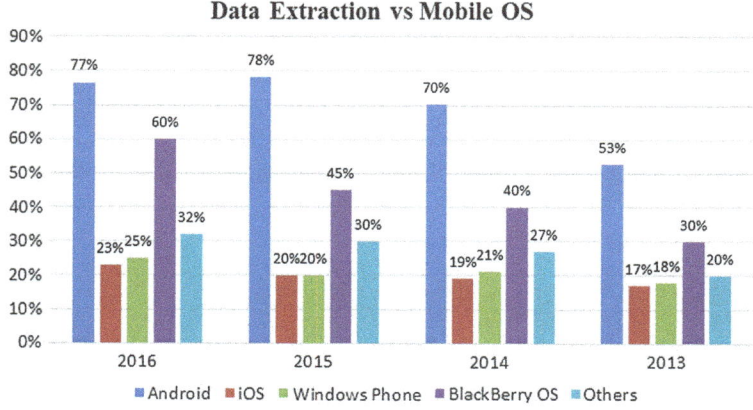

Fig. 1 Share of different Mobile OS. *Source* Gartner [1]

OS which covers more than 90–95% of the mobiles world over is Android, Windows and iOS. The term also includes forensics of Personal Digital Assistant (PDA), tablets and may include smart watches and various Internet of Things (IoT) devices in future, as all these digital devices can be considered mobile. One may argue that how can a smart TV be considered mobile. Well not by the literal term, but since it would be possible to have full control over any such IoT device remotely, a forensics analysis of the said device like any other mobile can be done in the times to come. Forensics includes carrying out analysis of the device to gather data from it both present and deleted. While in a few cases, the requirement may be to gather digital evidence to be produced in the Court of Law, and others may be used as a method to gather the source of malware infection in a large network. Forensic evidence extracted from mobiles can be an invaluable source of information for the investigator. A number of tools are used to extract valuable information from a given mobile phone. These tools can be commercial as well as open source. The present survey focuses on efficiency of such tools in extracting data and draws up a conclusion on how to take the research forward.

Figure 1 shows the share of different mobile OS which are used since 2013. As it is evident from the graph, the share of Android-based phones is on the rise.

Figure 2 specifies the different data acquisition processes. Maximum data can be extracted from physical method because it can extract deleted data also.

2 Literature Review

Numerous methods have been proposed by various authors to extract data from a smartphone which includes live and offline data. The tools available in the market have matured enough to capture information from the internal memory of the phone

Fig. 2 Data acquisition structure. *Source* [18]

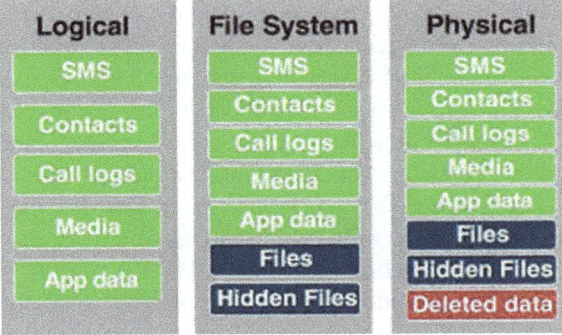

as well as external memory. Also, solutions towards extracting live data from the RAM of the phone have been proposed. The research work has been done for various types of operating systems which include Android, Windows and iOS in order of popularity among users. Lot of emphasis has been placed in recovering deleted data from various applications which are growing in number every day. As per the evaluation of different smartphone versions, various categories of data and mobile artefacts are increasing rapidly. Specific tools are used for different data extraction as per device and their OS specifications. While some tools are capable of physical extraction, others are able to get logical data only. Canlar et al. [3] had proposed a solution to carry out live forensics of a Windows mobile phone. This solution acquires data from both RAM and EEPROM of a Windows mobile phone using a SD card. The authors compare their solution Live SD with other tools (MIAT, instils, Paraben) to come to the conclusion that their method is unique and has the least memory footprint. Lohiya et al. [4] have discussed the various characteristics of mobile devices and the steps of acquiring digital evidence from a mobile phone. The hardware characteristics of a feature phone and a smartphone have been compared. The paper brings out the concepts of mobile forensics in general and lists a few forensic tools. However, comparison of a few basic artefacts of a Nokia E71 mobile phone using MobileditLite and Autopsy open-source tools only has been done. Mohtasebi et al. [5] studied trail version of four widely used mobile forensic tools namely Oxygen Forensic Suite, Paraben Device Seizure, Mobile Internal Acquisition Tool and Mobiledit Forensic and extracted data from a Nokia E5-00 smartphone. The technique of forensic acquisition using differential method has been introduced by Guido et al. [6]. The concept of hash comparisons is used to acquire the physical image from an Android phone. The authors have proved that their method is five times faster than any of the commercial tools available like UFED, XRY, Oxygen and the actual data can be acquired in less than 7 min. Wachter and Gruhn [7] discussed about acquisition and analysis of volatile memory of an Android smartphone. They have primarily concentrated on acquisition of volatile memory through Linux Memory Extractor (LiME) and analysis through volatility software, both of which are open source. The authors have also concluded that volatile data changes increase after 12 h from the start of the mobile phone and hence the importance of acquiring the volatile data as quickly as possible posts seizure. Authors Yang et al. [8] have proposed a physical acquisition method of an Android smartphone based on firmware update protocols. In order to update the OS or apply security patch in an Android phone, the firmware has to be updated. This feature has been exploited to access the flash memory of the phone where complete data is stored. A method to acquire live data as well as internal data of an Android phone has been proposed by Srivastava and Tapaswi [9]. The authors have given a method to get live RAM data along with logical acquisition of the Android mobile phone. Also, SD Card and eMMC storage imaging have been demonstrated by the authors. Thing et al. investigated [10], the dynamic behavior of the mobile phone's volatile memory and its usefulness in real-time evidence acquisition analysis of instant messaging applications. Kim et al. [11] have described how to acquire data from a CDMA cell phone through logical approach. A design for data acquisition has been proposed by the authors which is primarily using Embedded File System (EFS). The proposed method could extract phonebook, call history, SMS and photograph only. The authors Akarawita et al. [12] have proposed a framework to carry out forensic investigation of an Android smartphone. They have compared their work with a commercial tool like Oxygen Forensic Suit and an open-source tool ViaExtract CE and have arrived to a conclusion that their open-source tool is able to extract more data from the mobile (both physical and logical). The research proposed by Lessard and Kessler [13], discussed the general architecture of an Android device, its file system, etc. The research was carried out on a Sprint HTC Hero mobile running Android Ver. 1.5 which was the latest version available in the market in the year 2009. The data was also extracted using Universal Forensic Extraction Device (UFED) [14] device using a logical extraction methodology. The pros and cons of each method were listed out. While the experiment was done 7 years back, the methodology employed is still relevant for modern-day data extraction from smartphones.

All the papers discussed above describe existing methods using forensic tools to successfully extract data from different mobile devices. No one specific tool can be considered a universal one which can extract data completely. The efficiency and success in extraction of data are dependent on the OS version. Data extraction from the latest versions of the mobile is always a challenge.

3 Proposed Work

The literature survey done in Sect. 2 shows that data from a mobile can be extracted using various forensic tools which are both propriety and paid and those that have evolved from open source. However, there is no one universal method which can be used with 100% surety to fetch data from a mobile phone in a forensically sound manner. The present generation mobile forensic tools are not able to extract information from the latest mobiles due to a gap in the release of the newer version of the OS and the corresponding update of the forensic tool. This is a challenge which needs to be overcome. Non-availability of the phone architecture or the source code

Fig. 3 Areas of challenges
in mobile data extraction
Source [2]

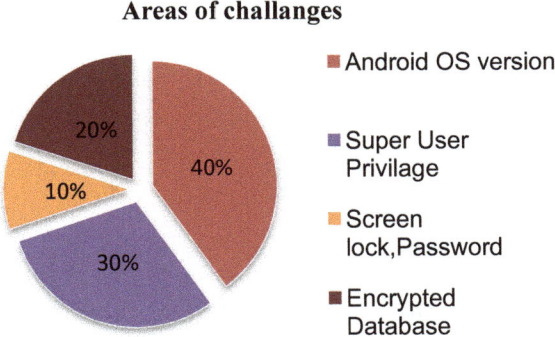

Areas of challanges

■ Android OS version

■ Super User
Privilage

■ Screen
lock,Password

■ Encrypted
Database

of the various apps, the increasing complexity of encryption of data both in rest and in motion and the increasing RAM size are hindering the recovery of data from the smartphones. There is no ISO-like standard as far as versioning of kernels of the phones is concerned or in the hardware architecture. This leads to failure of a solution developed for a particular model and version of kernel to be applied to other smartphones. Last but not least, the issue of pattern/pin lock and USB debugging enabling is always there [15].

Figure 3 shows the challenges faced at the time of data extraction for different version of Android phones. Multiple extraction tools are not able to extract complete data from the latest Android versions. Data extraction is easier if the device is rooted, and super user access is available ($SU). Bypassing screen lock and pattern lock is a challenge before getting root access on the device.

This research work proposes a solution of manual extraction of data from mobile in absence of any forensic tool. Results obtained exhibit that the manual extraction method proves to be at par in extracting the artefacts as compared to proprietary tools. However, automation of the process needs to be done through scripts to make it useful in a commercial setup.

4   Experimental Setup

In order to better understand the various data extraction techniques by different commercial data extraction tools (UFED, Paraben, XRY and Mobiledit), an experimental setup is being created in this research work. Study is being carried out on the forensic data obtained from different tools for different mobiles having different Android versions (v2.3.7, v4.0.5, v5) and comparative study of the results has been done. Subsequently, two newer versions of mobile (v5.0.1 and v6.0.1) were also taken which had been used extensively by a random user. These mobiles had actual user data and did not contain any pre-configured test data like the previous mobiles used for the study. Apart from extracting data through extraction tools, it is also tried

Table 1 Extraction of HTC Desire C (OS 4.0.5)

HTC Desire C (OS 4.0.5)	Tools used in extraction				
Extracted artefacts	UFED	XRY	Paraben	Mobiledit	Manual extraction
Contacts	29	29	29	29	29
SMS (**Present**)	07	28	06	05	04
SMS (*Deleted*)	04	00	00	00	00
Images	64	41	41	43	42
Web history	09	11	10	10	00
Call logs	00	05	00	02	00

to obtain the same information manually that does not involve any commercial or open-source tool. The idea was to get data from a mobile in the absence of a forensic tool. The results obtained from this could strengthen our belief that even those data can be extracted from a mobile which commercial tools fail to obtain. The manual approach [16] is using Android Debug Bridge (adb) commands which probably any commercial tool would be using in the background. These commands are popular android commands which are used both by app developers and security professionals to fetch and move data to and fro from a computer to an android device. A Linux-based virtual machine is used to carry out this task. This approach helped in refining the strategy and making the review more realistic.

5 Results and Discussion

The success of data recovery from mobiles is largely dependent on the available commercial tools. In the absence of these expensive commercial tools, the forensic investigator is unable to extract information from the mobiles. In order to obviate this problem, an exhaustive study was done to extract information from a mobile which included both deleted and that present in the phone using manual method. Based on the numerous experiments and the test results obtained thereof, the authors could safely conclude that the manual method was very powerful and could extract even those information which a commercial tool was unable to fetch. The result of test data obtained from two mobiles HTC desire C and Micromax A35 Bolt having Android versions 4.0.5 and 2.3.7 is as shown in Tables 1 and 2, respectively. Also, the results of data obtained from randomly picked mobiles LG Nexus 5 and Vivo Y51A versions 5.0.1 and 6.0.1 gave user data which is brought out in Tables 3 and 4, respectively. Non-rooting of mobiles prevented in carrying out a manual extraction for mobile is listed in Tables 3 and 4.

Table 2 Extraction of Micromax A35 Bolt (OS 2.3.7)

Micromax A35 Bolt (OS 2.3.7)	Tools used in extraction				
Extracted artefacts	UFED	XRY	Paraben	Mobiledit	Manual extraction
Contacts	30	30	30	30	30
SMS (**Present**)	07	28	06	05	04
SMS (*Deleted*)	04	00	00	00	00
Images	68	46	40	45	*109*
Web history	03	03	10	01	03
Call logs	18	18	0	18	18

Table 3 Extraction of LG Nexus 5 (OS 5.0.1)

LG Nexus 5 (6.0.1)	Tools used in extraction				
Extracted artefacts	UFED	XRY	Paraben	Mobiledit	Manual extraction
Contacts	521	123	–	–	–
SMS (**Present**)	30	24	–	–	–
SMS (*Deleted*)	06	00	–	–	–
Images	844	1244	–	–	–
Web history	135	120	–	–	–
Call logs	30	12	–	–	–

Table 4 Extraction of VIVO Y51A (OS 5.0.2)

VIVO Y51A (5.0.2)	Tools used in extraction				
Extracted artefacts	UFED	XRY	Paraben	Mobiledit	Manual extraction
Contacts	126	26	30	–	–
SMS (**Present**)	16	192	08	–	–
SMS (*Deleted*)	00	00	00	–	–
Images	2822	198	190	–	–
Web history	84	198	00	–	–
Call logs	884	41	05	–	–

Table 1 shows the data extraction summary of HTC Desire C Android mobile with respect to extracted artefacts like contacts, SMS, images, Web history and call logs. The number of artefacts extracted varied for each tool. For example, UFED [14] was the only tool which could extract deleted SMS and XRY [17] all present SMS.

Table 2 shows a similar extraction summary of Micromax A35 Bolt Android mobile with known artefacts as done in Table 1.

Similarly, Tables 3 and 4 demonstrate the extraction summary of LG Nexus 5 and Vivo Y51A mobiles which had actual user data. Wherever data could not be extracted due to non-rooted phone or non-support by the tool, those columns have been marked as '–'.

6 Conclusions

The present study has huge potential which can be expanded to develop universal solutions for different types of Android devices having different OS version. The proposed work has concentrated on Android phones which covers more than 84% of the mobiles in the world. The same concept can be extended to other operating systems like iOS, Windows and Blackberry and develop customized tools for these OS successfully. With the use of mobile growing exponentially and people shifting to m-commerce globally, detecting and preventing mobile crimes would be the need of the hour.

The following inferences can be drawn from the study:

- It is easy to extract nearly 100% data from older versions of Android due to immature technology in the past decade.
- Newer versions of Android pose varying degree of difficulty in extracting data.
- All data is present on the device till it is overwritten. Deleting data just resets the pointers but does not physically remove it from the memory. Thus, lot of data can be extracted from mobile.
- Manual extraction method is time-consuming and difficult as scripts have to be written for extracting each artefact. However, this method has the potential to extract more number of artefacts than any forensic tool.
- The novel method presented by the authors for data recovery from mobiles has the potential to obtain partially overwritten information, which can make the obtained evidence conclusive in a forensic investigation.

Acknowledgements The authors would like to express sincere gratitude to ITM University, Gwalior for providing the platform to work in cyber security as well as mobile forensics.

Disclaimer This study was not aimed at highlighting the efficiency or effectiveness of any particular commercial forensic tool. The research was done purely for academic purpose.

References

1. Market share alert: Preliminary, Mobile phones, Worldwide, 1Q17, May 2017
2. https://www.packtpub.com/books/content/introduction-mobile-forensics
3. Canlar S, Conti M, Crispo B, Di Pietro R (2013) Windows mobile LiveSD forensics. J Netw Comput Appl 36(2):677–684
4. Lohiya R, John P, Shah P (2015) Survey on mobile forensics. Int J Comput Appl 118(16)
5. Mohtasebi S, Dehghantanha A, Broujerdi HG (2011) Smartphone forensics: a case study with Nokia E5-00 mobile phone. Int J Digit Inf Wirel Commun (IJDIWC) 1(3):651–655
6. Guido M, Buttner J, Grover J (2016) Rapid differential forensic imaging of mobile devices. Digit Investig 18:S46–S54
7. Wächter P, Gruhn M (2015) Practicability study of android volatile memory forensic research. In: 2015 IEEE international workshop on information forensics and security (WIFS), 2015, pp 1–6
8. Yang SJ, Choi JH, Kim KB, Chang T (2015) New acquisition method based on firmware update protocols for android smartphones. Digit Investig 14:S68–S76
9. Srivastava H, Tapaswi S (2015) Logical acquisition and analysis of data from Android mobile devices. Inf Comput Secur 23(5):450–475
10. Thing VLL, Ng K-Y, Chang E-C (2010) Live memory forensics of mobile phones. Digit Investig 7:S74–S82
11. Kim K, Hong D, Chung K, Ryou J-C (2007) Data acquisition from cell phone using logical approach. In: Proceedings of the world academy of science, engineering and technology, vol 26
12. Akarawita IU, Perera AB, Atukorale A (2015) ANDROPHSY-forensic framework for Android. In: 2015 fifteen international conference on advance in ICT for engineering regions (ICTer), 2015, pp 250–258
13. Lessard J, Kessler G (2010) Android forensics: simplifying cell phone examinations
14. Cellebrite UFED Touch manual. www.mcsira.com
15. Hogg A. Android mobile forensic
16. Xda-devlopers (online) http://forum.xdadevelopers.com
17. MSAB.XRY—Extract: https://www.msab.com/products/XRY
18. http://www.advanceddiscovery.com/2016/10/11/mobile-device-discovery-challenges-part-4-acquisitions-and-afterwards/

Implementation of Image Compression and Cryptography on Fractal Images

Abhishek Madaan, Madhulika Bhatia and Madhurima Hooda

Abstract Fractal image compression can be implemented in many ways, such as IFS, affine transformations, and HV partitioning. Quadtree decomposition and Huffman encoding are used to improve compression. The work shows the existing methods like quadtree decomposition which is used to subdivide the original fractal image into unoverlapped blocks with a certain value of threshold. It also helps to improve compression ratio and peak signal-to-noise ratio. Huffman encoding is used with quadtree to decrease the number of bits required per symbol. The paper also involves the implementation of selective encryption using chaos on fractal images. A lot of data is being transferred every day, and that is why there is always a need of cryptography. Selective encryption is preferred over other encryption algorithms because it does not require to encrypt the whole image, i.e., image is only partially encrypted. Security analysis of proposed algorithm is also done in this paper which shows that this algorithm would be safe from any attack. MATLAB is used as a tool to implement the two algorithms.

Keywords Fractal images · Huffman encoding and decoding · Encryption
PSNR · Coefficient correlation

1 Introduction

Image storage and transmission is done on a daily basis. Its storage can be made more feasible using compression because many parts of an image have similarity and redundancy. In this paper, fractal images are used to implement compression.

A. Madaan (✉) · M. Hooda
Amity University, Noida, India
e-mail: abhishekmadaan717@gmail.com

M. Hooda
e-mail: 10madhurima@gmail.com

M. Bhatia
Manav Rachna International University, Faridabad, Haryana, India
e-mail: madhulikabhatia@gmail.com

© Springer Nature Singapore Pte Ltd. 2018
M. L. Kolhe et al. (eds.), *Advances in Data and Information Sciences*, Lecture Notes
in Networks and Systems 38, https://doi.org/10.1007/978-981-10-8360-0_5

Fractals are those objects which are used to describe nature occurring objects, and they exhibit similar patterns (they are self-similar). Examples of fractal images are lightening, river, space, blood vessels, etc. They can be modified using recursive algorithm on a computer. In fractal compression, image is subdivided into square blocks based on certain criteria, and then Huffman encoding is applied to this image to reconstruct and form a compressed image. Decoding phase of this algorithm returns a fine image and losses a very little data of the image. Fractal image compression is a lossy compression and to limit this loss, Huffman encoding is applied. In case of image transmission, security of that image is the most important part. That is why an encryption algorithm is also implemented on fractal images in this paper. Selective encryption using chaos is used to secure the image. It is a very efficient algorithm as it in encrypts only partially. The security of this algorithm is also analyzed in this paper using coefficient of correlation between the adjacent pixels, which proves that it is safe from any exhaustive search or other statistical attacks.

1.1 Fractals

One common question that arises is what fractal images are. Let us consider F as a fractal image, so F has following properties:

1. F is statistically self-similar (self-similarity means that an object is exactly or statistical similar).
2. F is nowhere differentiable according to mathematical equations.
3. Topological dimension of F is less than its fractal dimension (for instance, box dimension) [1].

Topological dimension of any fractal set F is n if neighborhoods of all points of that fractal have a boundary with $n - 1$ topological dimension, e.g., any set F has topological dimension 1 if its neighborhoods have 0 topological dimensions, while fractal dimension can be defined as the ratio which gives statistical index of complexity that how fractal's pattern changes with scale at which it is measured [2].

The box dimension of fractal set F is given by Eq. 1:

$$\log_{\epsilon \to 0} \frac{\log N_\epsilon(F)}{-\log \epsilon}, \tag{1}$$

where this limit exists.

Mathematical functions and contractive affine transformation are applied in the case of fractal image compression. Affine transformations are a combination of four transformations: scale, skew, translation, and rotation. Here, contractive means applying the transformation to the image brings any two points of original image closer and closer which limits it to a fixed point [1, 2].

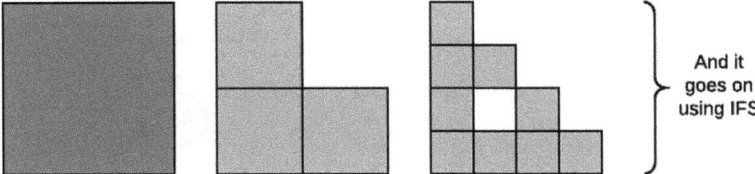

Fig. 1 IFS on an image

This fixed point is the resulting image. Another way of getting fractals is using iterated function system (IFS) as shown in Fig. 1. An IFS is a finite set of contractive mappings on a contractive metric system (CMS) [3].

2 Quadtree Decomposition

Nowadays, fractal geometry is an important field of nonlinear science and mathematics. Usually, two methods are used for fractal image compression: quadtree partitioning method and HV partitioning method [3].

Quadtree decomposition is an analysis technique in which the original image is subdivided into square blocks. These square blocks are more homogenous than the image itself. It helps to reveal the information about the images' structure. And it is the first step to initialize any image compression algorithm. It is done in the following way: A Quadtree is a tree data structure, so a quadtree partition is a representation of an image as a tree in which each node represents a square block of the image. Suppose if one wants to apply quadtree decomposition on a 128-by-128 intensity image, then the image is subdivided into 64-by-64 blocks. And then a certain criterion is tested on each block. Those blocks that fail to hold on the criterion are again subdivided into four descendants. This process continues until all the blocks meet this criterion [4].

And the criterion is given as follows: If the resulting optimal RMS value of one of the quadrants of the image is above the threshold value or if the depth of the country is less than the preselected maximum dimension, then range square is subdivided into four blocks [3, 4]. But these methods take too much time in encoding. Many techniques were developed to decrease the compression time but most of them have a bad effect on the fractal images after multiple iterations. To lessen the compression time and increase the compression ratio, hybrid encoding is done. It can be done by starting with quadtree decomposition and encoding with Huffman encoding algorithm [5].

Fig. 2 Huffman tree

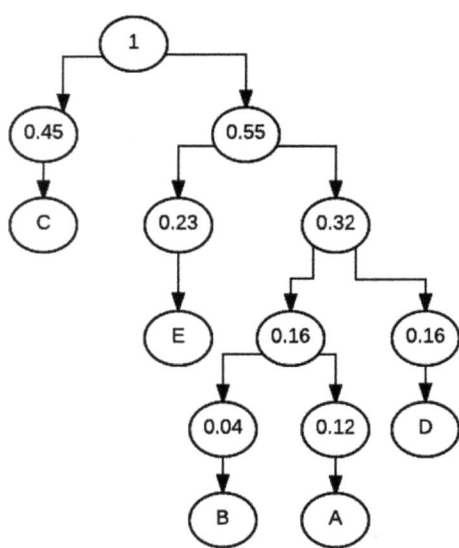

3 Huffman Coding

It was proposed by Dr. David A. Huffman, when he was a Ph.D. student, Electrical Engineering at MIT, in 1952. The idea behind the paper is to assign variable length codes to input alphabets/characters. Huffman codes are optimal. There are two steps to fulfill this coding:

(i) To make a binary tree from input character also known as Huffman tree.
(ii) Traverse that tree and assign prefix codes (variable length codes) to these characters [6].

(a) **Huffman Encoding (lossless image compression)**

It is a greedy algorithm. Firstly, it starts with constructing a list of alphabets with probabilities of frequencies of occurrence of these symbols.

A Huffman tree (as shown in Fig. 2) is constructed using following steps:

1. Create a leaf node for each alphabet/symbol and make a minimum heap of these nodes.
2. Take two nodes having the minimum frequency from this heap.
3. Create a new internal node whose frequency must be equal to the total of two node frequencies. Make the extracted node as its left descendant and the next node as its right descendant.
4. Repeat steps 2 and 3 until heap contains just one node, and it becomes a complete tree.

Then, represent right edge as one and left edge as zero. After this step, Huffman codes are determined for each symbol [4, 6].

Table 1 Symbols and their probability

Symbols	Probability	Bits/symbol	Code generated from tree	Length of each code generated
A	0.12	3	1101	4
B	0.04	3	1100	4
C	0.45	3	0	1
D	0.16	3	111	3
E	0.23	3	10	2

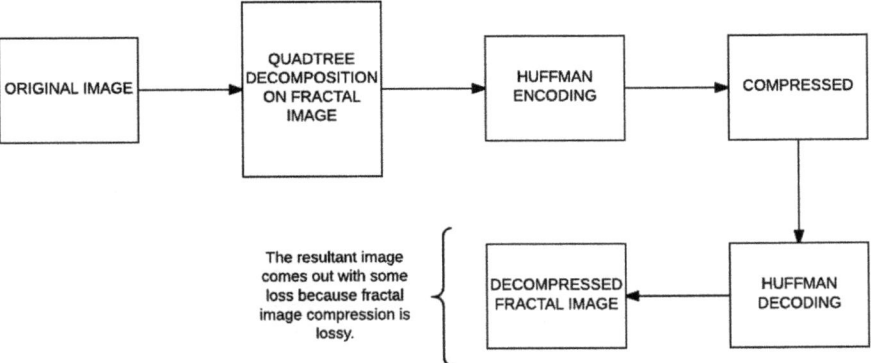

Fig. 3 Flow diagram for Quadtree decomposition and Huffman encoding

All these steps can be shown by using the following example shown in Table 1.

Efficiency of Huffman encoding: It can be shown from the above example that how Huffman encoding can be efficient. Before applying the encoding, each symbol takes up three bits. But, after applying it we can see the approximate improvement by multiplying probability and length of the symbol as follows:

$$(\text{Probability}) * (\text{length of each code generated})$$
$$= 0.12(4) + 0.04(4) + 0.45(1) + 0.16(3) + 0.23(2)$$
$$= 2.03 \text{ bits/symbol}$$
$$\text{Approx. Improvement} = \left(\frac{2.03}{3}\right) * 100 = 67.66\%$$

(b) **Huffman Decoding**

Huffman tree generally requires more bytes on the output side, but it is an efficient way to encode the image. The decoder must know that what is at the start of the encoded compressed file so that it can decode the whole input. Flow diagram for this algorithm is shown in Fig. 3.

Fig. 4 Original image

Fig. 5 Quadtree
decomposed

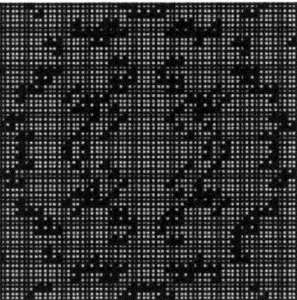

4 Proposed Approach

1. Divide the original image into square blocks using the values: minimum and maximum dimensions of that particular block must be 2 and 64, respectively, with threshold of 0.2. Command used in MATLAB: qtdecomp(A,0.2,[2 64]).
2. Store the block size, x and y coordinates, and mean value.
3. Apply Huffman encoding to this image to decrease compress the image effectively using commands 'Huffmandict,' 'huffmanenco.'
4. Compressed image is obtained.
5. Huffman decoding is applied on an encoded image to reconstruct that image and obtain decompressed image in return using command 'huffmandeco.' The results are shown in Figs. 4, 5 and 6 respectively.

4.1 PSNR

Peak signal-to-noise ratio is calculated to determine the difference between two images. It can be defined mathematically using Eqs. 2 and 3 [7]. The values calculated are shown in Tables 2 and 3 [1, 3].

Fig. 6 Decompressed image

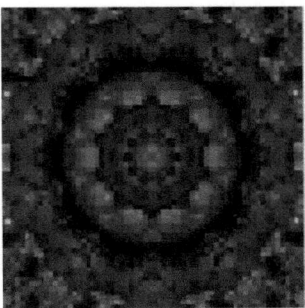

Table 2 Time taken for compression and decompression and compression ratio (only results using image $I1$ is used in this paper) [7, 13]

Image used	Time taken for compression (s)	Compression ratio	Time taken for decompression (s)
$I1$	1.83400	6.421164	12.17800
$I2$	1.68900	6.098712	18.99700

Table 3 MSE and PSNR value

Image used	MSE	PSNR value (in dB)
$I1$	6256.44	10.2015253
$I2$	11763.05	7.4595998

$$\text{MSE} = \sum_{M,N} \frac{[I1m, n - I2(m, n)]}{M * N} \tag{2}$$

$$\text{PSNR} = 10 \log_{10} \left(\frac{R^2}{\text{MSE}} \right), \tag{3}$$

where MSE is mean square error.

5 Necessity of the Use of Huffman Encoding

Huffman encoding is used here to increase the compression ratio. It can be seen in the example of Table 4 and its result that the approximate improvement comes out to be 67.66%. Quadtree decomposition for fractal images alone cannot improve the compression ratio as the decomposition of image into square blocks may result in distortion of the image at certain level.

Table 4 Entropy of original and encrypted images

Image used	Entropy of original image	Same for encrypted image
I1	6.96526	6.08551
I2	7.38991	6.04566

Fig. 7 Flow diagram for selective encryption using chaos

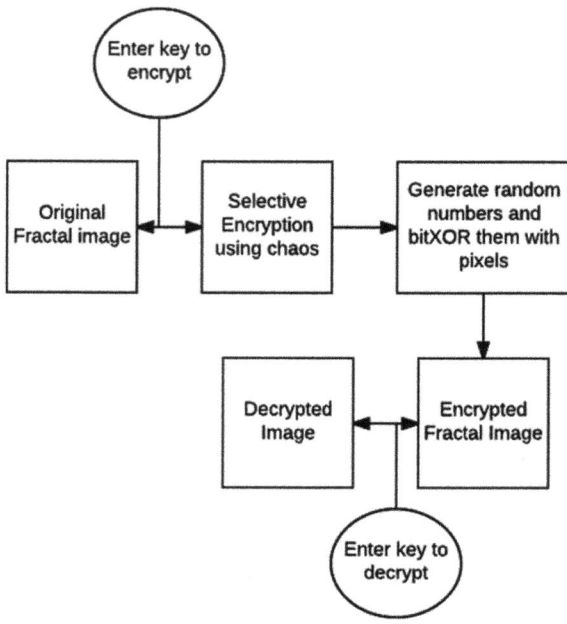

6 Selective Encryption

Cryptography is used to ensure authenticity and confidentiality of any image. It is one of the most important aspects of modern computing. Data must be encrypted before it is transmitted to make it secure from various attacks; there is a need for securing multimedia data, Internet communication, military communication, and medical imaging. Memory requirements for many algorithms like data encryption standard (DES) and advanced encryption standard (AES) are more than that for selective encryption [8]. Selective image encryption is used to encrypt only some content of the image. It helps to reduce the execution time and provides less overhead consumption. And it is also known as partial encryption. Consequently, this algorithm provides security to image, while some part of the image is still visible. It enables to achieve three major goals for image encryption, i.e., confidentiality, integrity, and availability. Selective encryption using chaos is used for colored images (RGB images) [8, 9]. Block diagram of this algorithm is shown in Fig. 7.

Concept of confusion and diffusion is used in this algorithm. Confusion means every pixel of the encrypted image should be dependent on several parts of the

Fig. 8 Original image

Fig. 9 Encrypted image

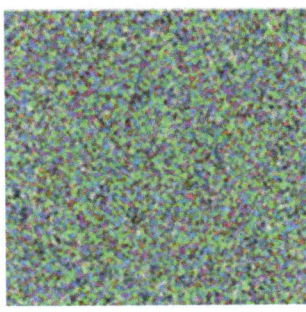

Fig. 10 Decrypted image

entered key. While diffusion means a little change in encrypted image would result in change in the bits of original image and vice versa [10]. Algorithm to apply selective encryption is as follows:

1. Input the original fractal image using imread.
2. Reduce its salt-and-pepper noise (or impulse noise) because it causes sudden and sharp disturbances in the image processing. Command used: medfilt2(i1,[3,3]);
3. Use the technique of confusion and diffusion to replace each pixel of the original image with a new value. This value can be generated by random number generator.
4. BitXOR those numbers with image pixels which results in an encrypted image. The results are shown in Figs. 8, 9, and 10, respectively.

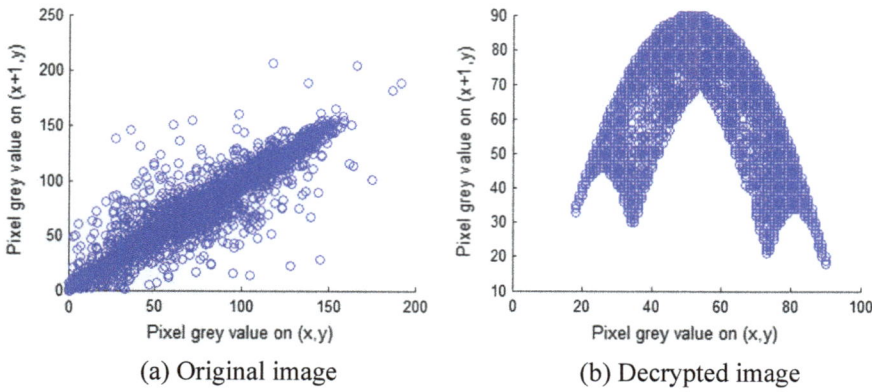

(a) Original image   (b) Decrypted image

Fig. 11 **a** Original image and **b** decrypted image

7 Security Analysis of Proposed Algorithm

An algorithm is secure only and only if it is safe from cryptanalytic threats, differential and exhaustive search attacks, etc. Statistical analysis is done by attackers such as permutation-based ciphers that can easily crack many cipher algorithms and recover original image from encrypted image. Therefore, it is necessary to confirm that this algorithm is safe from such attacks by using the concept of correlation coefficient [5, 11].

7.1 Correlation Coefficients

Correlation coefficient is calculated between adjacent pixels of original image and encrypted image. First step is to select several pairs of adjacent pixels randomly. The histogram in Fig. 11 clearly shows that there is an extraneous correlation between neighboring pixels of encrypted image while pixels in original image are well correlated. Hence, it proves that selection encryption using chaos is secure [5]. Then calculate using following formula:

$$C.C. = \frac{N(\sum X * Y) - (\sum X)(\sum Y)}{\sqrt{[N\sum X^2 - \sum X^2][N\sum Y^2 - (\sum Y)^2]}} \tag{4}$$

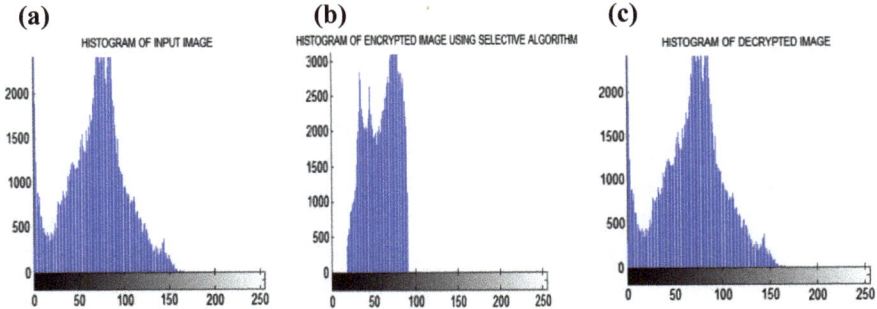

Fig. 12 **a** Histogram for original, **b** encryption, and **c** decrypted images

7.2 Information Entropy

It is a measure of randomness which is used to characterize the texture of an image statistically. In other words, it is used to determine a measure of pixels grayscale disturbances. It is defined in Eq. 5.

Table 4 shows the entropy values.

$$\text{Entropy, } H(x) = \sum_{i=1}^{n} pm_i \log_2 \frac{1}{p(m_i)} \tag{5}$$

where $m_i = m_1, m_2, \ldots, m_n =$ set of random variables, $p(m)$ is the histogram counts, and it can be obtained using imhist (inbuilt command in MATLAB) [9, 11, 12].

7.3 Histogram Analysis

The security of this algorithm can also be analyzed using histograms of original and encrypted image. According to the figure, it can be observed that the histogram of original image is a uniformly distributed histogram. While histogram of encrypted image is shuffled in such a way that attacker cannot compute the gray values of original image from encrypted image using statistical analysis. Hence, this encryption algorithm sustains any kind of exhaustive search and differential attacks [9]. Figure 12a–c shows the results.

8 Conclusion

After implementing these two algorithms, it can be concluded that fractal image compression using quadtree decomposition and Huffman encoding certainly helps to improve compression ratio and PSNR value. Image compression using only quadtree partitioning or HV partitioning does not give good results as they are lossy compressions. Two images are used to implement compression. After the analysis of these images in MATLAB, the following conclusions came up: PSNR of image 1 is 10.2015 dB and for another image, it is 7.4595 dB. Time taken for compression and decompression is also improved. Compression ratio after compressing image 1 is 6.4211 and for another image, it is 6.0987. There are many advantages to use fractal image compression, i.e., JPEG is six times slower than this algorithm, it provides better compression ratios than many algorithms, and it is resolution independent. But it also has some disadvantages; it takes much longer time to compress and decompress an image than JPEG, and it is little expensive for small projects. Image encryption using selective encryption with chaos reduces the amount of data to be encrypted. It is an efficient algorithm as it saves time to implement as it encrypts partially with the help of it can be concluded that coefficient of correlation between adjacent pixels for encrypted image is different from that of original image which implies that this algorithm is safe from any attack which ensures it safety, and histogram analysis shows that histogram for encrypted image is completely different from that of input image.

References

1. Nandi U, Mandal JK (2015) Fractal image compression with adaptive quadtree partitioning and archetype classification. In: 2015 IEEE international conference on research in computational intelligence and communication networks
2. Thamizhchelvy K, Geetha G A novel approach to generate fractal images using chaos theory. Indian J Comput Sci Eng (IJCSE)
3. Negi A, Agrawal A, Negi A (2014) A review on fractal image compression. Int J Comput Appl (0975–8887) 85(4)
4. Veenadevi SV, Ananth AG (2012) Fractal image compression using quadtree decomposition and Huffman coding. Signal Image Process Int J (SIPIJ) 3(2)
5. Bhatia M, Bansal A, Yadav D, Gupta P Proposed algorithm to blotch grey matter from tumored and non tumored brain MRI images. Indian J Sci Technol
6. Padmavati S, Mesharam V (2015) DCT combined with fractal quadtree decomposition and Huffman coding for image compression. In: 2015 international conference on condition assessment techniques in electrical systems (CATCON)
7. Pandey M, Bhatia M, Bansal A (2016) IRIS based human identification: analogizing and exploiting PSNR and MSE techniques using MATLAB. In: 2016 international conference on innovation and challenges in cyber security (ICICCS-INBUSH)
8. Thakur A, Kumar R, Bath A, Sharma J (2014) Design of selective encryption scheme using matlab. IJEEE 1(1)
9. Bisht U, Goswami S (2014) Analysis and implementation of selective image encryption technique using matlab. IOSR J Comput Eng (IOSR-JCE)
10. Meligy AM, Diab H, El-Danaf MS (2016) Chaos encryption algorithm using key generation from biometric images. Int J Comput Appl (0975–8887) 149(11)
11. Bhatia M, Bhatia M Object tracking in a video sequence using mean-shift based approach: an implementation using MATLAB7. IJCEM Int J Comput Eng Manag

12. Bansal A, Madhulika B, Implementing edge detection for detecting neurons from brain to identify emotions. Int J Comput Appl. Foundation of Computer Science, New York
13. Madhulika B, Yadav D, Gupta P, Kaur G, Singh J, Gandhi M, Singh A, Implementing edge detection for medical diagnosis of a bone in matlab. Comput Intell Commun Netw (CICN)

Computer-Aided Diagnosis of Melanoma Skin Cancer: A Review

Puneet Kumar Goyal, Nirvikar and Mradul Kumar Jain

Abstract Skin cancer has a major impact on society in India and across the world. According to the figures given by the National Cancer Institute and SEER, estimated new cases of Melanoma in 2017 are 87,110. This figure is approximated 5.2% of all new cancer in 2017. As per the data obtained from the WORLD HEALTH RANKINGS, the death rate per 1,00000 is highest in New Zealand with 7.68% then Australia with 6.52%. It has been proved from the study that melanoma skin cancer is almost curable if it is diagnosed early and treated correctly; otherwise, it can spread to other parts of the body and become incurable. This paper presents the comparative study of various phases of computer-aided melanoma skin cancer detection system with the aim of providing the development achieved in the melanoma skin cancer detection by the research community from earlier period to the current time. This method starts from the image acquisition step followed by image preprocessing, segmentation, feature extraction, feature selection and classification steps. The input to this system is an image of affected skin area, and output labels this input image benign or malignant melanoma.

Keywords Melanoma · Preprocessing · Segmentation · Classification · Benign
Malignant · Oncology · Epiluminescence microscopy · Neural network
Fuzzy C-means

P. K. Goyal (✉) · M. K. Jain
Uttarakhand Technical University, Dehradun, Uttarakhand, India
e-mail: puneet17jan@gmail.com

M. K. Jain
e-mail: mradul_ja@rediffmail.com

Nirvikar
COER Roorkee, Roorkee, Uttarakhand, India
e-mail: nirvikarlohan@yahoo.com

© Springer Nature Singapore Pte Ltd. 2018
M. L. Kolhe et al. (eds.), *Advances in Data and Information Sciences*, Lecture Notes
in Networks and Systems 38, https://doi.org/10.1007/978-981-10-8360-0_6

1 Introduction

Cancer is a major disease in which a group of abnormal cells in the body exhibits uncontrolled growth, while in a normal cell life cycle, the cells generate from other cells, grow and die when they are damaged. These abnormal cells are defined as cancer cells, malignant cell, or tumor cells. The severity of the cancer can be imagined from the fact that it can also spread to other parts of the body. For example, cancer cells in the skin can travel to the lung and grow up there. So this process of spreading cancer cells to other parts is called metastasis. There are many different types of cancer such as leukemia sarcoma, melanoma, and many more [1].

According to the American Cancer Society, an estimated 87,110 new cases of malignant melanoma will be identified in the USA in 2017 and expected 9,730 people will die due to melanoma skin cancer in 2017 [2]. In India, skin cancers comprise about 1–2% of all diagnosed cancers. Various types of skin cancers produce almost 2.4% of total cancer patients who are treated in the surgical oncology department. There are basically three types of skin cancer. Among these, squamous cell carcinoma (SCC) was the widespread histological type (55.8%) followed by melanoma (26.1%) and basal cell carcinoma (BCC, 18.1%).

Among all these skin cancers, melanoma is the deadly form of skin cancer [3]. Melanoma skin cancer arises when the pigment-producing cells (melanocytes) exhibit uncontrolled growth and become cancerous. Most pigment cells are found in the skin and generate pigments which provide color to the skin. It can also arise in the other parts of the body, such as the eyes, intestines (this is rare). It is very rare in people with dark skin [4, 5].

Skin cancer may appear as a new spot or mole which could be benign or malignant melanoma. Benign melanoma is harmless, while malignant melanoma is dangerous which needs immediate attention. If the melanoma skin cancer is not identified at starting phase, it could be the basis of death of the patient. As per the melanoma: statistics approved by the Cancer.Net Editorial Board, 07/2016, the 5-year survival is 92%, if it is diagnosed at initial stage (Fig. 1).

Basically, doctors use clinical analysis and biopsy process for the analysis of the skin cancer [2]. Clinical analysis is done using a dermatoscope by expert dermatologists [6]. Biopsy method is painful and time-consuming as it undergoes removal of the skin, and these skin samples are tested by many laboratories [7]. Also, the

Fig. 1 Melanoma [2]

accuracy of clinical analysis was 64%, while accuracy for biopsy method was 68% which is very low. Considering all these facts, a computer-aided detection (CAD) method is required which is capable of performing complex image processing and machine learning to diagnose the skin cancer efficiently and correctly.

2 Literature Categorization

This section below provides the related references and review on work done in the field of computer-aided detection of melanoma skin cancer. There are various steps for computer-aided detection of skin cancer such as

(a) Image acquisition
(b) Image preprocessing
(c) Image segmentation
(d) Feature extraction
(e) Feature selection
(f) Classification

2.1 *Image Acquisition*

It is one of the most important steps of CAD system for melanoma skin cancer detection. There are numerous image acquisition methods under investigation which are confocal scanning, laser microscopy, ultrasound, magnetic resonance imaging (MRI), optical coherence tomography (OCT), biopsy, etc. The usual clinical practice of melanoma diagnosis is a visual inspection by the dermatologist. Accuracy of clinical diagnostic is bit unsatisfactory. On the basis of various studies, we can say that the naked eye clinical diagnosis of cutaneous melanoma has an accuracy rate of only 60%.

Dermoscopy readers may refer to [8, 9] for performance analysis of existing image acquisition techniques. From the study of these research papers, it could be concluded that dermoscopy is the best suitable noninvasive method for image acquisition. There are various types of dermoscopy such as dermoscopy using non-polarized light that visualizes the subsurfaces of the PSL by using the microscope and immersion fluid [10–12] which makes the skin layer more transparent. This technique is also known as dermatoscopy, cutaneous surface microscopy, magnified oil immersion bioscopy, and epiluminescence microscopy (ELM).

To further improve the accuracy of image acquisition, a modified dermoscopy was conducted with cross-polarized light which eliminates the need of immersion liquid and direct contact of instrument with skin [10, 13]. This method is sometimes referred as video microscopy or XLM (X-polarized epiluminescence [10, 14]. Another image acquisition related to dermoscopy is transillumination technique

(TLM) which provides the better clarity of the image by directing the light on the skin in a manner so that back-scattered light illuminates the skin lesion. The device used for this method is called nevoscope [2, 15, 16].

Therefore, dermoscopy is referred to all techniques that provide the visualization of PSL by using surface microscopy [comp. analysis of pigmented skin].

In simple ELM, the diagnosis accuracy could be extended up to a certain limit. To further improve the clarity of the acquired image, digital dermoscopy analysis and D-ELM have been developed [17, 18]. But this digital evaluation of PSL requires sophisticated image processing software to help physicians in diagnosis process.

2.2 Preprocessing

In CAD system, the major challenge is to differentiate or detect lesion from the healthy skin as transition between the lesion and the surrounding skin is smooth. In digital dermoscopy, an image is taken in digital format to obtain clear image but it could be possible that image might have various artifacts such as hair, air bubbles, ruler markings, specular reflections, interlaced video field misalignment. These artifacts further degrade the quality of the acquired image and increase the chances of inaccurate detection of the lesion because there is very much similarity between the lesion and surrounding skin [19].

Therefore, preprocessing is the first step to improve the quality of the acquired images by removing the noise (unwanted signals) or artifacts such as hairs, bubbles.

If image preprocessing is not performed properly, then it might cause the inaccurate classification of an image [19] as well as increases the computation time.

There are various image preprocessing techniques for each artifact type. To remove hairs in dermoscopic image, dull Razor is used [20, 21], Kiani and Sharfat [21], Hoshyar et al. [22] improved it further to remove light-colored hairs. We can also apply various filters such as adaptive median filter, median filter, Gaussian filter, mean filter, Wiener filter, and adaptive Wiener filter for removing various noises such as Gaussian noise, salt-and-pepper noise, Poisson noise, and speckle noise [23].

In addition to further improve the image quality, some image enhancement methods are also used. In these methods, most important is color correction or calibration.

This method involves in recovery of real colors of a photographed lesion which is taken from low-cost digital camera [21, 24, 25]. Other approaches are illumination correction, contrast enhancement, and edge enhancement. To improve the contrast various methods like histogram stretching, histogram equalization, FFT, homomorphic filtering [26] could be used.

2.3 Segmentation

Image segmentation is the most important techniques after the preprocessing step. It is a process of continuously dividing an image into multiple parts until the region of interest (ROI) related to particular application has been detected. This step determines the eventual success or failure of the image analysis. There is a rare chance of failure of an efficient segmentation method. Here we have briefly provided an overview of various segmentation algorithms which are being used for dermoscopic image analysis as provided in

Method used	Advantages and disadvantages
Threshold-based segmentation It is the simplest image segmentation method. In this method, a single threshold value t is considered and the pixel located at coordinates(x, y) with grayscale value f is assigned to class 1 if $f \leq t$ or else the given pixel is allocated to class 2 The selection of the threshold depends on the value at which ROI becomes identified correctly. This method converts a grayscale image into binary image. Here are some algorithms of thresholding: (Otsu's method, local and global thresholding, maximum entropy, histogram based, etc.) [2, 19, 27–29]. Among all these segmentation algorithm, it has been concluded that Otsu algorithm gives optimum result	Advantages: • Computationally inexpensive • Fast and simple for implementation Disadvantages • Extremely susceptible to noise • The choice of correct threshold value is critical because incorrect choice may result in over- or under-segmentation
Region-based segmentation In this segmentation, the main concept is to classify or categorize a particular image into number of categories or regions. So we need to assign a class or category to which each pixel in the image belongs Region growing is a simple region-based segmentation method that merges or groups pixels by examining neighbor pixel of initial seed point and determine whether that pixel should be added to the group or not Region splitting and merging is another region-based segmentation method which divides an image into uniform regions. This algorithm starts by assuming that the entire image is a single region. Then calculate the homogeneity criteria if it is false, then divide the region into four smaller regions, and repeat this splitting until no further splitting is required [30]. Now, these small square regions are merged to form larger irregular regions. This process ends when no further merges are possible Some other famous algorithms for region-based segmentation are: (seeded region growing, watershed segmentation, etc.) [19, 27–29]	Advantages: • Gives better result in comparison with other segmentation methods • Proper selection of seed gives accurate result than any other methods Disadvantages: • Computationally expensive • Selection of noisy seed by user leads to flawed segmentation

Method used	Advantages and disadvantages
Fuzzy C-means method [19, 27–29]. It is one of the most popular unsupervised segmentation techniques. It is a method which categorizes one piece of data into two or more clusters. This method was developed by Dunn in 1973 and later enhanced by Bezdek in 1981. It is widely used in medical image segmentation like brain tumor detection, MRI.	Advantages: • FCM unsupervised and converge very well Disadvantages: • Sensitive to noise. Determination of fuzzy membership is not very easy
Edge detection approach [31, 32] It is a process of locating an edge in the selected image [33]. The edge representation of the image not only consists of significant information but also reduces the amount of data to be processed; therefore, this method is used by advance computer vision algorithms like medical image processing, biometric. The following steps are used for edge-based segmentation: Transform the original image into edge image which consists of all edges Now process the image to identify object boundaries Transform the result in simple segmented image [34] The edge detection method could be categorized into two classes such as Gradient and Laplacian The most frequently used edge detection techniques are Roberts edge detection, Sobel, Prewitt, Canny, Log edge detection, Marr–Hilderth edge detection [35]	Advantages: • This method is suitable for images which have good contrast among regions Disadvantages: • This method is not suitable for the images which have too many edges or ill-defined • It is less resistant to noise as compared to techniques
Neural network approaches [19, 27–29]. In neural network segmentation, an image is processed using ANN or a set of neural networks. In this method, small area of an image is given as input, and after performing the decision-making process, it marks the area of the input image into lesion or healthy tissues for medical image processing such as brain tumor, MRI.	Advantages: • There is no requirement to develop complex programs • This method uses the parallel processing capability of neural networks Disadvantages: • Training time is long; • Initialization may affect the result; • Overtraining should be avoided

2.4 Feature Extraction

In computer-aided detection of melanoma, features of skin image play an important role to determine whether it is benign or malignant melanoma. In the initial stage, both benign and malignant melanoma appear very much similar [2]. All these features are categorized as internal and external features. Internal features such as blue-white veil, cancerous area of skin, irregular streaks are extracted from dermoscopic image. External features are age of a person, family history related to cancer, itching on the skin, etc. [2]. Basically, which features should be extracted depend on the diagnosis method used for identifying melanoma. For example, asymmetry, border, color, and diameter are the features of the ABCD rule for melanoma skin cancer detection [21]. Therefore, for an effective computer-aided detection system, it is required to extract only those features which can be understood by a computer.

There are some conventional clinical diagnosis methods such as ABCD rule, Menzies method, and seven-point checklist. Among all these conventional methods, ABCD rule of dermoscopy works as the most effective method for many computerized melanoma detection systems due to its implementation simplicity [36]. After pigmented skin lesion was determined, the features extracted from affected skin lesion and its surrounded normal skin area are divided into color based, border based, symmetry and texture based [37]. Rahil et al. [36] proposed an effective feature extraction method which combines texture and border features extracted from pigmented skin lesion. Abdul et al. [19] used a graylevel co-occurrence matrix (GLCM) in which contrast, correlation, homogeneity, and energy features are used with three additional features related to geometry of the image. Therefore, feature extraction creates new features from functions of the original features.

2.5 Feature Selection

In machine learning, feature selection is the process of selecting a subset of relevant features (variables, predictors) for use in model construction. Basically, this step is done after feature extraction and before classification step. It is a technique which reduces dimensions widely used for efficient data mining and knowledge discovery. But it is required to ensure that important information should be preserved during dimensionality reduction.

For developing the method of feature selection, different researchers have taken different approaches [11]. A very useful review on feature descriptors was given by Maglogiannis and Doukas [4] in 2009. Others methods are leave-one-out (LOO), sequential floating forward selection (SFFS), and sequential floating backward selection (SFBS), considered for this purpose. Ganster et al. [38] used SBFS and SFFS with subset size between 10 and 15 features. But the performance of this method degrades for more than 20 features.

According to the study, melanoma recognition methods perform well for small subsets followed by a slight increase up to medium-sized subsets [12], while performance of melanoma identification decreases for larger subsets. This thing was confirmed by Ruiz et al. [14] using SBFS and SFFS evaluation and discovered that minimum error rate was achieved using subset of six features and a significant increase in classification error rate is observed for subset of more than 20 features.

Yashar et al. [39] defined a particle swarm optimization–support vector machines (PSO–SVM) feature selection method that reduced the number of features effectively and chosen the best subset for their purpose. PSO computationally is less expensive than other methods.

2.6 Classification

Lesion classification is the final step in computerized analysis of melanoma skin cancer detection. This step is used to decide whether the skin lesion is malignant or benign. To perform the classification task, the existing systems use different classification methods to the features that were extracted in prior stage. There exist some different classifiers such as logistic regression, discriminant analysis, artificial neural network, K-nearest neighborhood, support vector machine, decision trees, and support vector machine (SVM) [15]. The classification performance is evaluated w.r.t. classification accuracy, sensitivity, specificity, positive predictive value (PPV), negative predictive value (NPV), likelihood ratio, etc.

Year	Classification method [Source]	Sensitivity (%)	Specificity (%)	Accuracy (%)
2011	kNN classifier [40] Ramlakhan et al. [40] used the OpenCV kNN classifier to categorize skin lesion as benign or malignant. In this classifier, learning process is based on training instances. The class to which test sample belongs is based on the majority of its k-most similar instances of training set	60.7	80.5	66.7
2012	Multilayer perceptron [41] Mariam A. Sheha et al. used the MLP which is a three-layer feedforward neural network. It uses two techniques for classification which are automatic MLP and traditional MLP. In first one, given data are categorized into three subsets which are training, validation, and testing to perform classification, while in second one, complete data are used for training	70.5	87.5	76
		92.3	91.6	92
2015	SVM classifier with RelisfF filter-based method [42] Luis Rosado et al. [42] used the SVM classifier with ReliefF filter by using the features of asymmetry criterion, border, and color criterion	86	73	76.7
2015	The classification is performed by the SVM in two stages. In the first, classifier is constructed in training stage, and then classifier performance is done by using tests which are independent of training set [43]	95	83.33	90.63

3   Conclusion

Early detection of melanoma skin cancer plays an important role to decrease its death rate drastically. This paper has discussed all the phases of computer-aided melanoma skin cancer detection method in detail. From the review presented in this paper, it has been concluded that there are many algorithms for melanoma skin cancer detection starting from the painful laboratory testing to the computer-aided detection system, and among these algorithms, SVM classification gives more accuracy and sensitivity than kNN classifier and multilayer perceptron. Therefore in future to eliminate the drawbacks of these algorithms such as pain or problem that a patient feels during laboratory testing and to increase the accuracy of melanoma detection, the computer-aided diagnosis (CAD) by using some advanced algorithms like deep learning, machine learning are used.

References

1. Lee H, Chen YPP (2015) Image based computer aided diagnosis system for cancer detection. Expert Syst Appl 42:5356–5365
2. Mehta P, Shah B (2016) Review on techniques and steps of computer aided skin cancer diagnosis. In: International conference on computational modeling and security, CMS 2016
3. Cancer scenario in India: as per the statistics. Posted on 2/11/2014 by Daily excelsior
4. Geller AC, Swetter SM, Brooks K, Demierre M, Yaroch AL (2007) Screening, early detection, and trends for melanoma: current status (2000–2006) and future directions. J Am Acad Dermatol 57:555–572
5. Maglogiannis I, Doukas CN (2009) Overview of advanced computer vision systems for skin lesions characterization. IEEE Trans Inf Technol Biomed 13(5):721–733
6. Garnavi R, Aldeen M, Bailey J (2012) Computer-aided diagnosis of melanoma using border and wavelet-based texture analysis. IEEE Tran Inf Technol Biomed 16(6):1239–1252
7. Vestergaard ME, Menzies SW (2008) Automated diagnostic instruments for cutaneous melanoma, pp 1085–5629. https://doi.org/10.1016/j.sder.2008.01.001
8. Rigel DS, Russak J, Friedman R (2010) The evolution of melanoma diagnosis: 25 years beyond the ABCDs. Wiley Online Library 60(5):301–316
9. Nakariyakul S, Casasent DP (2008) Improved forward floating selection algorithm for feature subset selection. In: International conference on wavelet analysis and pattern recognition, 30–31 Aug 2008, vol 2, pp 793–798
10. Madooei A, Drew MS (2013) A colour palette for automatic detection of blue-white veil. IEEE 2013
11. Celebi ME, Aslandogan YA (2004) Content-based image retrieval incorporating models of human perception. In: Proceedings of international conference on information technology: coding and computing (ITCC), vol 2, IEEE Computer Society Press, Los Alamitos, CA, p 2415
12. Zhou H, Chen M, Zou L, Gass R, Ferris L, Drogowski L, Rehg JM (2008) Spatially constrained segmentation of dermoscopy images. In: 5th IEEE international symposium on biomedical imaging, pp 800–803
13. Liu H, Yu L (2005) Toward integrating feature selection algorithms for classification and clustering. IEEE Trans Knowl Data Eng 17:491–502
14. Ruiz D, Berenguer V, Soriano A, SáNchez B (2011) A decision support system for the diagnosis of melanoma: a comparative approach. Expert Syst Appl 38(12):15217–15223

15. Sumithra R, Suhil M, Guru DS (2015) Segmentation and classification of skin lesions for disease diagnosis. In: International conference on advanced computing technologies and applications (ICACTA-2015). Procedia Comput Sci 45:76–85

16. Celebi ME, Kingravi HA, Uddin B, Iyatomi H, Aslandogan YA, Stoecker WV, Moss RH (2007) A methodological approach to the classification of dermoscopy images. Comput Med Imaging Graph 31(6):362–373

17. Sasikala M, Kumaravel N (2005) Comparison of feature selection techniques for detection of malignant tumor in brain images. In: IEEE Indicon 2005 conference, Chennai, India, 11–1 3 Dec 2005

18. Nakariyakul S, Casasent DP (2008) Improved forward floating selection algorithm for feature subset selection. In: Proceedings of the 2008 international conference on wavelet analysis and pattern recognition, Hong Kong, 30–31 Aug 2008

19. Jaleel JA, Salim S, Aswin RB (2013) Computer aided detection of skin cancer. In: International conference on circuits, power and computing technologies (ICCPCT), IEEE 2013

20. Lee T, Ng V, Gallagher R, Coldman A, McLean D (1997) DullRazor: a software approach to hair removal from images. Comput Biol Med 27(6):533–543

21. Korotkov K, Garcia R (2012) Computerized analysis of pigmented skin lesions: a review. Artif Intell Med 56(2):69–90

22. Kiani K, Sharafat AR (2011) E-shaver: an improved DullRazor® for digitally removing dark and light-colored hairs in dermoscopic images. Comput Biol Med 41(3):139–145

23. Hoshyar AN, Al-Jumaily A, Hoshyar AN (2014) Comparing the performance of various filters on skin cancer images. In: International conference on robot PRIDE 2013–2014, Published by Elsevier

24. Quintana J, Garcia R, Neumann L (2011) A novel method for color correction in epilumines-cence microscopy. Comput Med Imaging Graph 35(7–8):646–652

25. Wighton P, Lee TK, Lui H, McLean D, Atkins MS (2011) Chromatic aberration correc-tion: an enhancement to the calibration of low-cost digital dermoscopes. Skin Res Technol 17(3):339–347

26. Maglogiannis I, Zafiropoulos E, Kyranoudis C (2006) Intelligent segmentation and classifica-tion of pigmented skin lesions in dermatological images. https://doi.org/10.1007/11752912_23. In: DBLP conference: advances in artificial intelligence, proceedings of 4th Helenic con-ference on AI, SETN 2006, Heraklion, Crete, Greece, 18–20 May 2006

27. Ebrahimi SM, Pourghassem H, Ashourian M (2010) Lesion detection in dermoscopy images using Sarsa reinforcement algorithm. In: 17th Iranian conference IEEE biomedical engineering (ICBME)

28. Ali AR, Couceiro MS, Hassenian AE (2014) Melanoma detection using fuzzy C-means clus-tering coupled with mathematical morphology. In: 14th international conference IEEE hybrid intelligent systems (HIS)

29. Sookpotharom S (2009) Border detection of skin lesion images based on fuzzy C-means thresh-olding. In: 3rd international conference IEEE genetic and evolutionary computing. WGEC'09

30. Splitting and Merging, [Online]. Available at http://homepages.inf.ed.ac.uk/rbf/CVonline/LOCAL_COPIES/MARBLE/medium/segment/split.htm

31. Delgado D, Butakoff C, Ersboll BK, Stoecker WV (2008) Independent histogram pursuit for segmentation of skin lesions. IEEE Trans Biomed Eng 55:157–161

32. Zhou H, Chen M, Zou L, Gass R, Ferris L, Drogowski L, Rehg JM (2008) Spatially constrained segmentation of dermoscopy images. In: 5th IEEE international symposium on biomedical imaging, pp 800–803

33. Chapter 3. Review of image segmentation methods. Available at http://shodhganga.inflibnet.ac.in/bitstream/10603/9107/8/08_chapter3.pdf

34. Chapter 10. http://www.cs.uu.nl/docs/vakken/ibv/reader/chapter10.pdf

35. Muthukrishnan R, Radha M (2011) Edge detection techniques for image segmentation. Int J Compu Sci Inf Technol 3(6)

36. Garnavi R, Aldeen M, Bailey J (2012) Computer-aided diagnosis of melanoma using border and wavelet-based texture analysis. IEEE Trans Inf Technol Biomed 16(6):1239–1252

37. Ikuma Y, Iyatomi H (2013) Production of the grounds for melanoma classification using adaptive fuzzy inference neural network. In: IEEE international conference systems, man, and cybernetics (SMC)
38. Ganster H, Pinz P, Rohrer R, Wildling E, Binder M, Kittler H (2001) Automated melanoma recognition. IEEE Trans Med Imaging 20(3)
39. Maali Y, Al-Jumaily A (2012) Hierarchical parallel PSO-SVM Based subject-independent sleep apnea classification. ICONIP 2012, Part IV, LNCS 7666, pp 500–507
40. Ramlakhan K, Shang Y (2011) A mobile automated skin lesion classification system. In: Proceedings of the 23rd IEEE international conference on tools with artificial intelligence (ICTAI'11), pp 138–141
41. Mabrouk MS, Sheha MA, Sharawy A (2012) Automatic detection of melanoma skin cancer using texture analysis. Int J Comput Appl 42(20):22–26
42. Rosado L, João M, Vasconcelos M, Ferreira M (2015) Pigmented skin lesion computerized analysis via mobile devices. SCCG 2015, Smolenice, Slovakia, ACM, 22–24 April 2015. http://dx.doi.org/10.1145/2788539.2788553
43. Correa DN, Paniagua LR, Noguera JL, Pinto-Roa DP, Toledo LA (2015) Computerized diagnosis of melanocytic lesions based on the ABCD method. In: Computing conference (CLEI), 2015 Latin American, 19–23 Oct 2015, pp 1–12

Computer Vision-Based Tomato Grading and Sorting

Sukhpreet Kaur, Akshay Girdhar and Jasmeen Gill

Abstract Since ages, agricultural sector plays an important role in the economic development of a country. In recent years, industries have started using automated systems instead of manual techniques for quality evaluation. In agriculture field, grading is very necessary to increase the productivity of the vegetable products. Everyday a huge amount of vegetables are exported to other places and earn a good profit. So, quality evaluation is important in terms of improving the quality of vegetables and gaining profit. Traditionally, the vegetable grading and classification were done through manual procedures which were error prone and costly. Computer vision-based systems provide us such accurate and reliable results that are not possible with human graders/experts. This paper presents a vegetable grading and sorting system based on computer vision and image processing. For this work, tomatoes have been used as a sample vegetable. A total of 53 images were acquired using own camera setup. Afterward, segmentation using Otsu's method was performed so as to separate the vegetable from the background. The segmented images, thus obtained, were used to extract color and shape features. At last, grading and sorting were performed using backpropagation neural network. The proposed method has shown an accuracy of 92% and outperformed the existing system.

Keywords Computer vision · Image processing · Grading and sorting
Feature extraction · Artificial neural network

S. Kaur (✉) · A. Girdhar
Guru Nanak Dev Engineering College, Ludhiana, India
e-mail: sukhpreet885@gmail.com

J. Gill
RIMT IET, Mandi Gobindgarh, India

© Springer Nature Singapore Pte Ltd. 2018   75
M. L. Kolhe et al. (eds.), *Advances in Data and Information Sciences*, Lecture Notes
in Networks and Systems 38, https://doi.org/10.1007/978-981-10-8360-0_7

1 Introduction

Agriculture plays a significant role in the economic development of a nation. In developing countries like India, 70% people depend on agriculture. It is a major source of income for the country. It is considered as the backbone of Indian economy [1]. Grading of vegetables after harvesting is very important. Customers are getting more quality conscious, so it is very necessary to grade the vegetables for the quality purpose. The grading of vegetables is done on the basis of physical properties like size, color, shape, weight, and specific gravity. The well-known methods for grading are manual grading and size grading. When vegetables arrive at the processing centers in a supermarket, they are graded for quality concern. The immature, mature, and over-mature vegetables are sorted and then graded on the basis of maturity level.

In recent years, development in an agriculture field has been possible by using computer vision techniques. But still, development is very slow as compared with other development fields like electronics and automobile. Therefore, new methods should be developed to forefront agriculture field. There are still old methods that are used for quality evaluation of agriculture products. The quality inspection is very important to increase the market value of products. Therefore, some new automated methods are required for quality checking which are more accurate and reliable than old methods [2].

Nowadays, quality inspection is done by using traditional methods which are not so easy. With the development of image processing tools and computer vision techniques, it becomes very easy to evaluate the quality of vegetables. As the existing methods have the drawback of low speed, high complexity, high cost, and low efficiency, so it is necessary to develop a system for vegetable grading and sorting with low cost and high speed [3]. The agriculture products follow many different steps like cutting of products, washing, sorting, grading, packing, transporting, and then storing. Among all these steps, sorting and grading are the main steps to find out the good quality products. Sorting is done on the basis of physical appearance, and grading totally depends upon the quality of vegetables. A number of feature attributes that are taken for grading are size, shape, color, texture, etc.

A robot has been deployed in a different type of fields like automobile, consumer goods, medical, aerospace, food, agriculture, and many other fields. The robot helps to do task accurately with computer vision techniques. These tasks are pre-defined. Computer vision system has many applications like object recognition, performing agriculture tasks, remote sensing, face recognition, biometrics, and medical image analysis. Weed detection, identification of medical plants, disease detection, classification of grains, classification of crops and weeds are the various agriculture applications. In these systems, image analysis is carried out to obtain the feature set that is used for classification and grading [4].

Humans use their brain and eyes to see the world around them. Computer vision is the similar technique that provides same capability to the machine of human level recognition. It is a technique for acquiring, processing, analyzing, understanding, and extracting meaningful information from images to produce numerical and symbolic data [5]. It is a technology for developing artificial intelligence systems that gain information from images and multi-dimensional data. The importance of artificial intelligence is to develop a system which acts as a human being and can perform mechanical actions.

The rest of the paper is organized as follows: Sect. 2 discusses the related work, Sect. 3 discusses the proposed method, and Sect. 4 discusses the experimental results. Finally, Sect. 5 concludes the paper.

2 Related Work

This section contains various algorithms that have been proposed by different authors for classification and grading of agriculture produces.

Pla et al. [6] presented a sorting system based on the maturity degree, skin defects, size, and weight. Classification was done on the basis of size and color. It can process 15 fruits per second and sort then according to the extracted parameters. Chalidabhongse et al. [7] developed a vision-based mango's sorting system which can extract 2-D or 3-D visual parameters of mangoes. The basic parameters that were extracted were size, area, surface area, and volume. Firstly, the silhouette part of the mangoes was extracted using segmentation. All the necessary features were then extracted, and backpropagation neural network was used for sorting the mangoes.

Feng and Qixin [8] presented a fruit sorting system. Classification feature was fruit's color ratio that was calculated with HIS color space. The classic Bayes classifier was used for sorting the fruits. The average accuracy rate of 90% was achieved. Mustafa et al. [9] developed a grading and sorting system of agriculture produces. These products were classified based on size and shape parameters using support vector machine, and grading was done using a fuzzy logic technique.

Nandi et al. [10] presented a machine vision maturity prediction system for sorting mangoes. Initially, 27 features were extracted and then most relevant features were selected. Support vector machine was used as a classifier to predict the maturity level. The average accuracy of the system was 96%. Ukirade [11] presented a method based on color grading of tomatoes for quality detection. In this work, color features were taken to classify tomatoes, as color was the deciding factor of maturity. A total of 20 images were used to train the network. The tomatoes were graded into four maturity levels for red, orange, turning, and green with the grades A, B, C, and D, respectively. A backpropagation neural network classifier was used to test the maturity level.

Saito et al. [12] developed an eggplant grading system. Color and shape parameters were extracted, and eggplants were classified based on the artificial neural network. Calpe et al. [13] presented a fruit sorting and grading system. In this system, degree of ripeness was measured and decisions were made on that measured parameters. Lee et al. [14] developed an automated fruit quality evaluation system. Quality and maturity of the fruits were measured by using color components. In this, 3-D RGB components were converted into linear color index and results were more promising.

Pavithra et al. [15] developed a machine vision-based cherry tomato sorting system. In this work, color-based maturity estimation algorithm was used and grading was done on the basis of color, texture, and shape parameters. K-nearest neighbor and support vector machine classifiers were used to classify matured tomatoes. Bhatt [16] described an apple classification system. The proposed system was based on machine vision and artificial neural network which classified the apple on the basis of size, color, and external defects. It shows overall accuracy of 96%.

Mhaski et al. [17] developed a system for checking the quality of tomatoes. Size and shape and degree of ripeness parameters were used for inspecting the tomatoes. K-mean clustering classifier is used for quality evaluation of tomatoes. Lee et al. [18] presented a color mapping concept for fruit quality evaluation. 3-D RGB components were mapped into 1-D space. The quality of the fruits was checked using direct color mapping. Mizushima and Lu [19] presented an automatic segmented method for apple grading and sorting. This work presented segmentation of color images using Otsu's thresholding algorithm and support vector machine (SVM).

All the existing methods were capable of classifying and grading the agriculture products. Color, shape, texture, and size features were most commonly used in these methods. In existing methods, grading of tomatoes was done on the basis of maturity level. Defected or diseased tomatoes were not considered for grading. Moreover, choosing a threshold value for segmentation was also a challenging task. Therefore, any future research endeavors may look into these challenges and such algorithms are required to be developed that provide more reliable solutions. The objective of this research method is to grade tomatoes-based quality which must be nondestructive. For this backpropagation, neural network (BPNN) algorithm is used which is relatively simple and standard method. It is the classic gradient-based algorithm to find the best weight vectors and minimize the cost function.

3   Proposed Method

In existing work, a method was developed for quality detection which was based on maturity of tomatoes. In this, color features were taken to classify the tomatoes, as color was the deciding factor of maturity. A total of 20 images were used to train the network. The tomatoes were graded into four maturity levels for red, orange, turning, and green with the grades A, B, C, D, respectively. A backpropagation neural network classifier was used to test the maturity level.

In proposed method, 'tomato' has been used as a sample vegetable. Firstly, images are acquired and database of images is prepared. Then, the images are segmented to separate the region of interest from the background details. Afterward, color and shape features of tomatoes are extracted. At last, extracted features are fed as input to the backpropagation neural network classifier for grading and sorting of tomato. The following steps are followed in order to grade the tomatoes, and the corresponding flowchart is shown in Fig. 1.

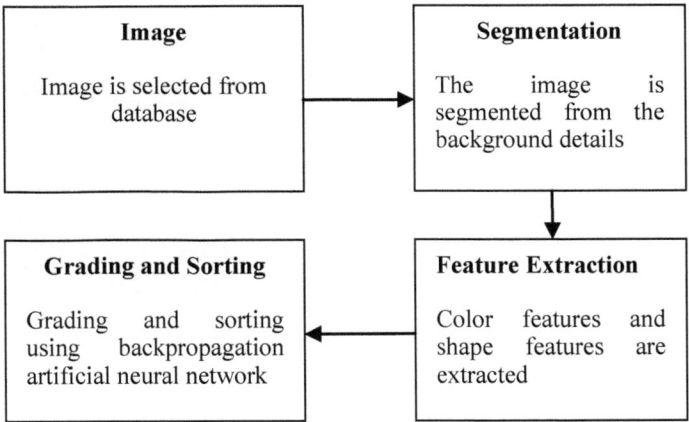

Fig. 1 Flowchart of the grading and sorting process

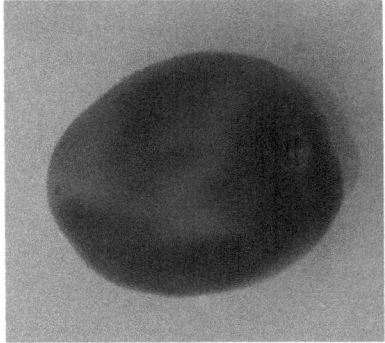

Fig. 2 Grayscale image

A. Segmentation

Image segmentation is used to separate an image into multiple parts. The main aim is to change the representation into something which is easier to analyze. In this, the image is converted into grayscale image, and then the segmentation is done by using Otsu's method of thresholding (see Figs. 2 and 3). A threshold value is obtained, and segmentation is done to extract the region of interest. Finally, a binary image is generated.

B. Feature Extraction

In this, the set of features that can be useful in grading and sorting are selected. The color and shape features are extracted, and a feature set is generated. Color features are obtained from Red Green Blue (RGB) components. RGB components of sample images are separated and then, the mean and standard deviation of RGB components are calculated. Shape features are calculated using a segmented image.

Fig. 3 Segmented image

C. Backpropagation Neural Network (BPNN)

Backpropagation is used to grade the tomatoes. It is a supervised learning algorithm and a layered feed-forward artificial neural network that is basically used to train the network. It calculates the errors with respect to assigned weights and optimizes the performance of the network by adjusting the weights. The error is the difference between the actual output and the desired output. Basically, it is used to minimize the error [11].

4   Results and Discussion

A. The Experimental Results

This section presents the experiments and quality grading results of the proposed system. Image processing toolbox in MATLAB R2014a (Mathworks, MA) has been used for simulation. The evaluation of proposed approach has been performed on a single machine with Intel (R) Core (TM) i5-4200 M CPU @ 2.5 GHz, 4 GB of RAM, and Windows 8.0 Pro as an operating system.

The two types of tomatoes are used, defected and non-defected. The tomatoes are graded into two grades. Grade A is for non-defected tomatoes and grade B is for defected tomatoes. The results of the two grades are shown in Fig. 4. A total of 53 images are captured in which 28 are used for training and 25 are used for testing the system.

Results are based on different parameters like color and shape of the tomatoes. Various color features like mean and standard deviation are calculated from RGB components of the image. The min and max values of color features are shown in Table 1. Along with color features and shape features like area, major axis length, minor axis length, eccentricity, convex area, solidity, and perimeter are also calculated as shown in Table 2. Afterward, a feed-forward BPNN is used to grade the tomatoes. A total of 13 features (6 color and 7 shape features) are fed as input to BPNN and 20 hidden layers are used and one output is generated. Each input value is multiplied by

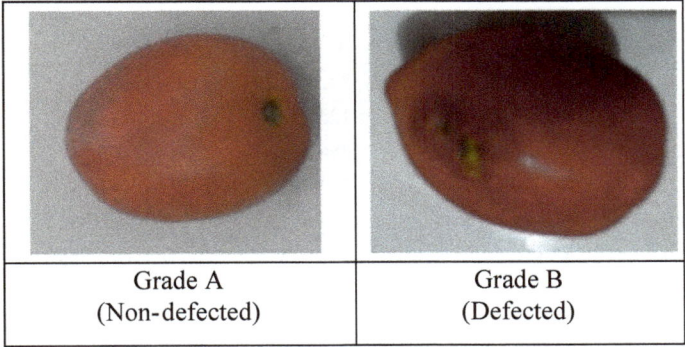

Grade A (Non-defected)	Grade B (Defected)

Fig. 4 Grading category for tomatoes

Table 1 Stores mean and standard deviation of RGB components

Grade				A	B
Mean	R	Min		165.1334	115.9402
		Max		174.2684	152.5644
	G	Min		130.5480	78.5692
		Max		155.6730	101.1196
	B	Min		119.4552	65.2261
		Max		138.7248	88.6473
Standard deviation	R	Min		15.2400	31.3633
		Max		18.1459	49.5666
	G	Min		29.1225	64.2360
		Max		54.2688	76.2710
	B	Min		51.5554	73.4680
		Max		64.2262	83.7465

the connection weight then summation of these multiplied values is calculated and passed through a function called transfer function or activation function to generate the desired output.

B. Performance evaluation

For accomplishing this, the accuracy performance evaluator is used. This evaluator includes the components like True Positive (TP), True Negative (TN), False Positive (FP), and False Negative (FN). TP shows correct positive prediction, TN shows correct negative prediction, FP shows incorrect positive prediction, and FN shows incorrect negative prediction.

Accuracy is the most common measure determined from the confusion matrix. Accuracy of the classifier is determined by Eq. (1).

Table 2 Stores shape features

Grade		A	B
Area	Min	37,195	85,164
	Max	123,724	117,896
Major axis length	Min	386.3985	661.5452
	Max	783.4960	761.7418
Minor axis length	Min	357.8715	450.9845
	Max	701.0133	676.5635
Eccentricity	Min	0.1543	0.2745
	Max	0.7665	0.7738
Convex area	Min	74,688	186,364
	Max	247,292	240,124
Solidity	Min	0.4240	0.4311
	Max	0.5003	0.4997
Perimeter	Min	2339.91	3557.16
	Max	4670.91	4526.48

Table 3 2 × 2 confusion matrix of proposed work

2 × 2 confusion matrix		Predicted grade		Accuracy (%)
		A	B	
Original grade	A	12	1	92.3
	B	1	11	91.6

$$\text{Accuracy} = \frac{TP + TN}{TP + TN + FP + FN} \tag{1}$$

Confusion matrix of proposed work is a 2 × 2 matrix for Grade A and Grade B. In this matrix, columns represent the instances of predicted class and rows represent the actual class (or vice versa). The accuracy of 'A' grade tomatoes is 92% and of 'B' grade tomatoes is 91%. The average accuracy of the system is 92%. Confusion matrix of proposed work is shown in Table 3.

Table 4 represents the confusion matrix of existing work [11]. In existing work, tomatoes dataset is taken and tomatoes are graded into four maturity levels for red, orange, turning, and green with the grades A, B, C, D, respectively. Testing is done with total 40 tomatoes in which 15 are of 'A' grade, 7 are of 'B' grade, 10 are of 'C' grade, and 8 are of 'D' grade. The average accuracy of the existing system is 90%.

From the graph shown in Fig. 5, it is clear that the proposed method produces better results in terms of accuracy (92% accuracy) as compared to existing method which produced only 90% accuracy.

Table 4 4 × 4 confusion matrix of existing work [11]

4 × 4 confusion matrix		Predicted grade				Accuracy (%)
		A	B	C	D	
Original grade	A	14	1	0	0	93
	B	1	6	0	0	85
	C	0	1	8	1	80
	D	0	0	0	8	100

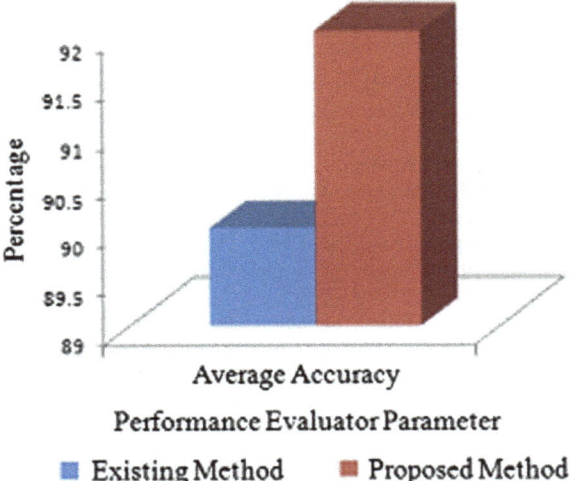

Fig. 5 Comparison of existing and proposed method

5 Conclusion and Future Scope

This paper presents a computer vision-based tomato grading and sorting that can be used to provide better accuracy under different circumstances. In this method, a methodology was developed to sort and grade the vegetables based on the color and shape features. The dataset contains samples of tomatoes. Six color features and seven shape features were extracted from each image. Feed-forward back propagation artificial neural networks (BPNN) were used as a classifier to sort and grade the tomatoes. In the neural network architecture, 20 hidden layers had been used for generation and movement of weighage to different samples. In proposed system, one 'A' grade tomato was misclassified by grade 'B' and one 'B' grade tomato was misclassified by grade 'A.' The proposed system outperformed the existing system as the accuracy of the proposed system is 92%.

The images captured are taken by using a white background. It can be modified to provide output for any kind of background. Also, calyx part was not removed from the tomatoes. This can be further extended using calyx elimination module.

References

1. Pandey R, Naik S, Marfatia R (2013) Image processing and machine learning for automated fruit grading system: a technical review. Int J Comput Appl 81(16):29–39
2. Sharma R, Kaur M (2015) ANN based technique for vegetable quality detection. IOSR J Electron Commun Eng 10(5):62–70
3. Ramprabhu J, Nandhini S (2014) Enhanced technique for sorting and grading the fruit quality using MSP430 controller. Int J Adv Eng Technol 7(5):1483–1488
4. Danti A, Madgi M, Anami BS (2014) A neural network based recognition and classification of commonly used indian non-leafy vegetables. Int J Image, Graph Signal Process 10:62–68
5. Kumar A, Gill GS (2015) Automatic fruit grading and classification system using computer vision: a review. In: Proceedings of advances in computing and communication engineering, Dehradun, pp 598–603
6. Pla F, Sanchiz JM, Sanchez JS (2001) An integral automation of industrial fruit and vegetable sorting by machine vision. In: Proceedings of emerging technologies and factory automation, France, pp 541–546
7. Chalidabhongse T, Yimyam P, Sirisomboon P (2006) 2D/3D vision-based mango's feature extraction and sorting. In: Proceedings of the international conference on control, automation, robotics and vision, Singapore
8. Feng G, Qixin C (2004) Study on color image processing based intelligent fruit sorting system. In: Proceedings of international conference on intelligent control and automation, China, pp 4802–4805
9. Mustafa NBA, Ahmed SK, Ali Z (2009) Agricultural produce sorting and grading using support vector machines and fuzzy logic. In: Proceedings of international conference on signal and image processing applications, Malaysia, pp 391–396
10. Nandi CS, Tudu B, Koley C (2014) A machine vision-based maturity prediction system for sorting of harvested mangoes. IEEE Trans Instrum Measure 62(7):1721–1730
11. Ukirade NS (2014) Color grading system for evaluating tomato maturity. Int J Res Manage Sci Technol 2(1):41–45
12. Saito Y, Hatanaka T, Uosaki K (2003) Eggplant classification using artificial neural network. In: Proceedings of the international joint conference, Portland, pp 1013–1018
13. Calpe J, Pla F, Monfort J (2002) Robust low-cost vision system for fruit grading. In: Proceedings of electrotechnical conference, Bari, pp 1710–1713
14. Lee DJ, Chang Y, Archibald JK (2008) Color quantization and image analysis for automated fruit quality evaluation. In: Proceedings of international conference on automation science and engineering, Arlington, pp 194–199
15. Pavithra V, Pounroja R, SathyaBama B (2015) Machine vision based automatic sorting of cherry tomatoes. In: Proceedings of electronics and communication systems, Coimbatore, pp 271–275
16. Bhatt AK (2013) Automatic apple grading model development based on back propagation neural network and machine vision, and its performance evaluation. AI Soc 30(1):45–56
17. Mhaski RR, Chopade PB, Dale MP (2015) Determination of ripeness and grading of tomato using image analysis on Raspberry Pi. In: Proceedings of international conference on communication, control and intelligent systems, Mathura, pp 214–220
18. Lee DJ, Archibald JK, Xiong G (2011) Rapid color grading for fruit quality evaluation using direct color mapping. IEEE Trans Autom Sci Eng 8(2):292–302
19. Mizushima A, Lu R (2013) An image segmentation method for apple sorting and grading. Comput Electron Agric 94:29–37

Movie Recommendation System Using Genome Tags and Content-Based Filtering

Syed M. Ali, Gopal K. Nayak, Rakesh K. Lenka and Rabindra K. Barik

Abstract Recommendation system has become of utmost importance during the last decade. It is due to the fact that a good recommender system can help assist people in their decision-making process on the daily basis. When it comes to movie, collaborative recommendation tries to assist the users by using help of similar type of users or movies from their common historical ratings. Genre is one of the major meta tag used to classify similar type of movies, as these genre are binary in nature they might not be the best way to recommend. In this paper, a hybrid model is proposed which utilizes genomic tags of movie coupled with the content-based filtering to recommend similar movies. It uses principal component analysis (PCA) and Pearson correlation techniques to reduce the tags which are redundant and show low proportion of variance, hence reducing the computation complexity. Initial results prove that genomic tags give the better result in terms of finding similar type of movies, and give more accurate and personalized recommendation as compared to existing models.

Keywords Movie recommendation · Genome tags · Content-based filtering
Vector space

S. M. Ali · G. K. Nayak · R. K. Lenka (✉)
IIIT Bhubaneswar, Bhubaneswar, India
e-mail: rakeshkumar@iiit-bh.ac.in

S. M. Ali
e-mail: syedmohdali121@gmail.com

G. K. Nayak
e-mail: gopal@iiit-bh.ac.in

R. K. Barik
KIIT University, Bhubaneswar, India
e-mail: rabindra.mnnit@gmail.com

© Springer Nature Singapore Pte Ltd. 2018
M. L. Kolhe et al. (eds.), *Advances in Data and Information Sciences*, Lecture Notes
in Networks and Systems 38, https://doi.org/10.1007/978-981-10-8360-0_8

1   Introduction

With the rapid development of Internet, data is growing at a very high pace, having said so many online movie platforms are exploding with new content everyday. Recommendation systems have proved to be one of the successful information filtering system. In general, recommendation systems are used to predict how much user may like a certain product/service, compose a list of N best items for user, and compose a list of N users for a product/service. With this growth of media, users have to spend significant amount of time searching for the movies in which they are interested [1]. Here, the task of recommendation system is to automatically suggest users what movie to watch next based to the current movie and thus saving their time searching for the related content.

Movie recommendation is the most used application on the media streaming Web sites, both in academics and as well as commercially research has been done extensively in this topic. The Netflix Prize challenge [2] is one such example where a prize of one million dollar was at stake. The aim of the competition was to beat the Netflix's very own recommender system by ten percent. This attracted various researchers and companies and more than forty thousand entries were submitted for the same. Most of these recommender use collaborative filtering mechanism which has been developed in recent few years [3–5]. First the ratings of movies are collected given by each individual and then the recommendation is given to the users based on similar type of people with similar taste in the past. Many popular online services like netflix.com, youtube.com have used this collaborative filtering technique to suggest media to users.

This work is an attempt at implementation of a recommender system which uses genome tags to find out similar types of movies and then basic content-based filtering technique to further enhance the results. MovieLens dataset [6] is used for the development of this recommender engine and is accessed through its public FTP interface [7].

2   Related Work

2.1   Recommendation Using Collaborative Filtering

Collaborative filtering recommender relies heavily on the users data or some contribution in order to make recommendation. Contributions may include users to give ratings, like, dislikes, or other kinds of feedback which can cluster similar type of users together. As the name suggest, is a way of recommendation for a user in "collaboration" with other users. The fundamental of this filtering lies in the fact that if a person A shares same interest as of B, for certain object(here movie), then A will more likely share the same interest as of B, for a different object, than that of a randomly chosen person [8]. Collaborative filtering is easy to implement, it works well in most of the cases and has ability to find links between items which are otherwise considered dissimilar [9]. One of the drawbacks that this system suffers from "cold start" problem, this happens when either a new user comes or new item is added and we do not have much information/feedback for the item/user [10, 19]. To overcome this problem, recommender user content-based recommendation techniques coupled with collaborative recommendation.

2.2 Recommendation Using Content-Based Filtering

Recommendation is purely made on the attribute of the item, hence avoiding "cold start" problem. The attribute of an item, for example, in a movie can be its genre, year, running time, rating, starring actors can be used by the content-based recommender to make a movie recommendation. This concept has its root from the information retrieval theory where a document representation methods can abstractly encapsulate features of an item for potential recommendation [11, 20].

Over the period of time content-based recommender starts building profile for a person, it includes the taste of an individual which is extremely helpful for highly personalized recommendation [3]. This type of recommender does not require community data as it solely relies on the individual's preference, hence explanation can be given why a certain item/media was recommended. One major disadvantage of this is that it requires contents which can be broken down into meaningful attributes.

3 Proposed Approach

The dataset which was downloaded for the research contained 24404096 ratings and 668953 tag applications across 40110 movies. These data were created by 259137 users between January 09, 1995 and October 17, 2016. And was generated on October 18, 2016. The dataset contained the following files "genome-scores.csv", "genome-tags.csv", "links.csv", "movies.csv", "ratings.csv", and "tags.csv". Description of who's is as follows

- genome-scores.csv: Contains the genome score of the movies corresponding to the tags.
- genome-tags.csv: Contains genome tag id and its corresponding string.
- links.csv: Contains the link to the other sources of movie data.
- movies.csv: Contains information about movie like its title, movie id, and genres.
- ratings.csv: Each row of this file contains rating of one movie by one user.
- tags.csv: Each row of this file represents one tag applied to one movie by one user.

3.1 The Genome Tags

Netfilx and Youtube are using hybrid of collaborative and content-based filtering, the prominent feature of which is genre of the video or movie. The major problem with the genre is that they are binary in nature, i.e., they do not tell till what extent that genre applies to the certain content. A user may apply the tag 'violent' to 'Fight Club', indicating that it is a violent movie, but they might not indicate how violent the movie is. Just like in Human Genome Project where all the genes in human DNA were identified and mapped, the researchers were inspired to find and index the building blocks of their media. Pandora has developed their own Music Genome Project [12], similarly for movie recommendation GroupLens research laboratory developed tag genome [13]. The genome tag extends the traditional tagging system to give the enhanced user interaction. Genome tag contains item and its relationship to the set of tags. These range between 0 and 1, where 1 being the most relevant and 0 being the least. This creates a dense matrix in which every movie in the genome has a value for every tag. This can be used to recommend similar type of content.

3.2 Data Preparation

Since genome score does not consist all the movies present in the dataset, first task was to select only those movies who's genome score we have. After filtering, we were left with around 10,000 movies. Next we transformed genome score which was stored like in Table 1 to like in Table 2.

After the transformation of genome scores average rating and number of users who rated the particular movie was calculated. This will come in handy while coupling our model with content-based filtering. Average rating for a particular movie i was calculated simply using the formula, total rating given by each user to that movie divided by the total number of users rated that movie, i.e.,

$$avg\_rating_i = \frac{\Sigma user\_rating_i}{\Sigma user_i} \tag{1}$$

Table 1 Before transformation

MovieId	TagId	Relevance
1	1	0.02400
1	2	0.02400
1	3	0.05475
1	4	0.09200

Table 2 After transformation

MovieId/tagId	1	2	3	...	1128
1	0.024	0.024	0.0548	...	0.0252
2	0.038	0.0418	0.037	...	0.0202
3	0.042	0.0525	0.0272	...	0.0200
4	0.036	0.0385	0.035	...	0.0140

3.3 Feature Reduction

We have total of 1128 genome tags, which are very large and many of them are redundant, this will increase the computational complexity. There is need to reduce the number of features, it will not only remove redundancy in data but will also increase the performance of the model [14]. Principal component analysis (PCA) was run on the complete genome score in order to the variance explained in data by the tags available [15].

As it is quite clear from the Fig. 1 that most of the tags show very low variance in data and can be reduced. We used correlation-based feature selection technique, for which Pearson correlation method was preferred [16]. Pearson correlation find out linear correlation between two variable X and Y. It results in the value between −1 and +1 where −1 is the total negative correlation and +1 being the total positive. Correlation between all the tags were calculated, if the tags were to be correlated they should have a value between 0 and +1. We were suppose to choose the optimal threshold value above which we will say tags are correlated else not, lets say this to be cutoff value. PCA was ran after selecting the cutoff to be 0.6, 0.5, 0.4, and 0.3 (Figs. 2 and 3).

Choosing cutoff to be 0.6 and 0.5 will still leave some of the redundant tags, whereas if cutoff is set to be 0.3 we might loose some important tags, so for this

Fig. 1 PCA on compete set of tags

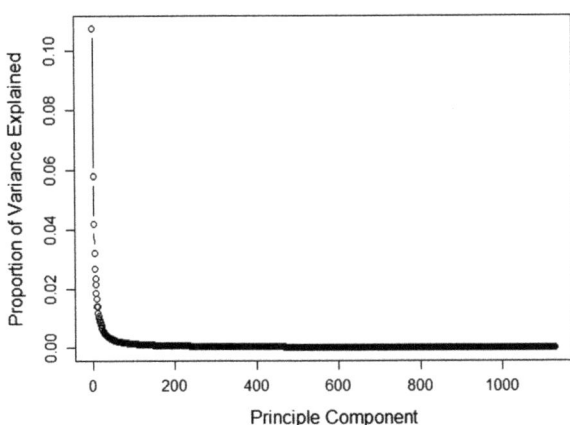

Fig. 2 PCA after cutoff was 0.6

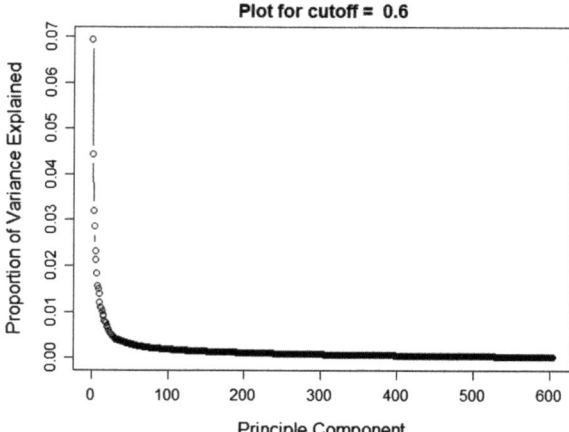

Fig. 3 PCA after cutoff was 0.5

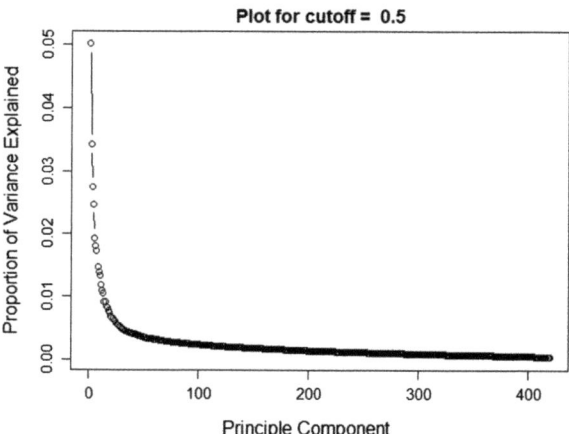

Fig. 4 PCA after cutoff was 0.4

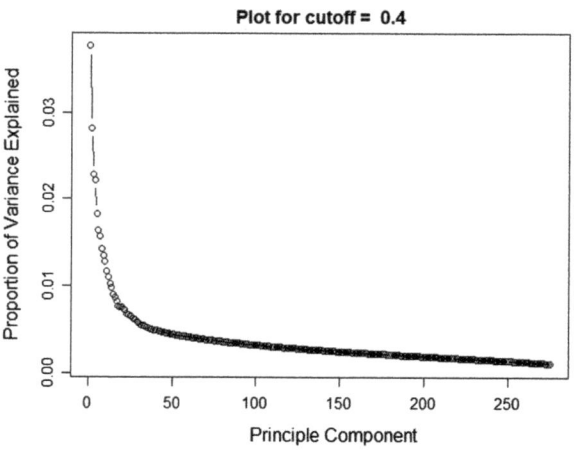

Fig. 5 PCA after cutoff was 0.3

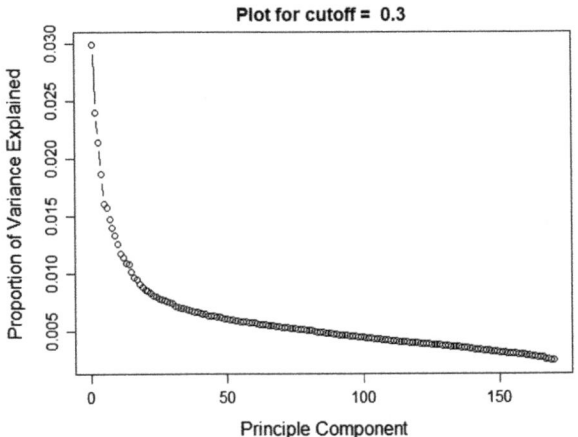

model cutoff was set to be 0.4. With this cutoff, the number of tags was dramatically reduced from 1128 to 275 (Figs. 4 and 5).

3.4 Distance Between Movies

Now we are left with 275 attributes and over 10,000 rows, we can use this to find out which movies are similar to whom. Vector space model approach was used to achieve this task [17]. In vector space model, we represent documents (or any object in general, here movie) as vectors of identifier. Relevance between these vectors can be find by comparing the deviation between then angles of all the vectors [18]. In practice, cosine of the angle between vector is calculated (Fig. 6).

$$\cos(\theta) = \frac{a_1 \cdot a_2}{\|a_1\| \|a_2\|} \tag{2}$$

where $a_1 \cdot a_2$ is the dot product.

$$\|a_1\| = \sqrt{\sum_{i=1}^{n} q_i^2} \tag{3}$$

In our dataset, every row can be treated as a vector, hence we can find out the cosine distance between each of them. For the demonstration purpose, we have chosen sample 2000 movies in this experiment. We obtained the 2000 * 2000 matrix M, where M_{ij} will contain the value of the cosine angle between the $movieId_i$ and $movieId_j$ (Fig. 7).

	A	B	C	D	E	F	G	H	I	J	K	L	M	N	O	P	Q	R	S	T	U	V	W
1	movid	50954	51007	51024	51037	51063	51077	51080	51084	51086	51088	51091	51094	51174	51255	51277	51357	51372	51412	51418	51471	51520	51540
2	50954	1	0.498753	0.495805	0.451851	0.396126	0.435289	0.502954	0.636487	0.437689	0.438088	0.554736	0.661334	0.578493	0.517129	0.456038	0.397745	0.481782	0.475393	0.501641	0.409672	0.738982	0.419267	0.
3	51007	0.498753	1	0.528159	0.478424	0.425265	0.398565	0.579962	0.517883	0.409785	0.410442	0.549749	0.524909	0.616765	0.595783	0.536569	0.46157	0.528474	0.423381	0.582499	0.532603	0.520383	0.47306	0.
4	51024	0.495805	0.528159	1	0.548621	0.583709	0.488179	0.649979	0.573352	0.529499	0.52986	0.532852	0.538495	0.632765	0.571545	0.615931	0.614043	0.621443	0.537956	0.531726	0.651673	0.572231	0.559605	0.
5	51037	0.451851	0.478424	0.548621	1	0.404146	0.407984	0.566059	0.484079	0.584026	0.421894	0.500871	0.481189	0.528304	0.546561	0.470418	0.532357	0.536305	0.606748	0.506773	0.429000	0.483887	0.507284	0.
6	51063	0.396126	0.425265	0.583709	0.404146	1	0.387449	0.477386	0.486409	0.417193	0.591676	0.404467	0.478374	0.526273	0.476818	0.431956	0.431564	0.528716	0.441478	0.460507	0.361705	0.4646	0.422853	0.
7	51077	0.435289	0.398565	0.488179	0.407984	0.387449	1	0.459874	0.485246	0.433395	0.385614	0.464676	0.470206	0.542498	0.494824	0.432778	0.347044	0.461938	0.506795	0.441924	0.360887	0.539379	0.422118	0.
8	51080	0.502954	0.579962	0.649979	0.566059	0.477386	0.459874	1	0.587148	0.55458	0.493877	0.606629	0.544804	0.705647	0.626531	0.577171	0.786891	0.618374	0.567248	0.620723	0.716888	0.533867	0.77613	0.
9	51084	0.636487	0.517883	0.573352	0.484079	0.486409	0.485246	0.587148	1	0.489225	0.480066	0.582305	0.684745	0.704743	0.613942	0.467308	0.426696	0.62268	0.558213	0.577229	0.51052	0.605751	0.498816	0.
10	51086	0.437689	0.409785	0.529499	0.584026	0.417193	0.453395	0.55458	0.489225	1	0.467011	0.474824	0.532189	0.578823	0.49948	0.420507	0.469407	0.506847	0.582831	0.470023	0.384515	0.450769	0.512692	0.
11	51088	0.438688	0.410442	0.52986	0.421394	0.591676	0.385614	0.493877	0.490066	0.467011	1	0.512904	0.547947	0.556528	0.623639	0.458937	0.472859	0.539207	0.469233	0.450851	0.373055	0.457298	0.445756	0.
12	51091	0.554736	0.549749	0.532852	0.500871	0.404467	0.464676	0.606629	0.582305	0.474824	0.512904	1	0.617058	0.704230	0.684029	0.355602	0.492087	0.596876	0.499898	0.604714	0.442313	0.565478	0.309005	0.
13	51094	0.661324	0.524909	0.538495	0.481189	0.478374	0.470206	0.544804	0.684745	0.532189	0.547947	0.617058	1	0.705114	0.574496	0.471205	0.424067	0.559395	0.532428	0.543339	0.44015	0.638637	0.508965	0.
14	51174	0.578493	0.616765	0.632765	0.528304	0.526273	0.542498	0.705647	0.704743	0.578823	0.556528	0.704230	0.705114	1	0.70629	0.574661	0.565699	0.630364	0.614565	0.651785	0.529543	0.65135	0.638639	0.
15	51255	0.517129	0.595783	0.571545	0.546561	0.476818	0.494824	0.626531	0.613942	0.49946	0.623639	0.684029	0.70629	1	0.602911	0.517289	0.639624	0.539807	0.683429	0.522643	0.565074	0.530094	0.
16	51277	0.456038	0.536569	0.615931	0.470418	0.431956	0.432778	0.577171	0.467308	0.420507	0.458937	0.355603	0.574661	0.602911	1	0.513393	0.538779	0.453703	0.562748	0.481839	0.500409	0.491246	0.

Fig. 6 Cosine angle between movies

```
> recommend_movie(51540)
[1] "Recommendation for movie "
[1] "51540 Zodiac (2007)"
# A tibble: 5 × 5
  movieId                        title                     genres avg_rating no_of_user
*   <int>                        <chr>                      <chr>      <dbl>      <int>
1   51080                Breach (2007)             Drama|Thriller   3.598284       1399
2  103624 Fruitvale Station (2013)                      Drama   3.612457        289
3   98961 Zero Dark Thirty (2012) Action|Drama|Thriller   3.704366       2405
4   82378 All Good Things (2010) Drama|Mystery|Thriller   3.055556        261
5   62577 Flash of Genius (2008)                      Drama   3.265432        162
> |
```

Fig. 7 Recommendation for "Zodiac"

3.5 Recommending Movies

Now when we have obtained the cosine matrix, the recommendation is fairly easy from here. Suppose a user watching a movie with movieId n, then we will go to the nth of the matrix and sort out the row in decreasing order. Greater the cosine value, smaller the angle, smaller the angle more close the movies are in vector space, more closer the movie more similar it is. We will pick up top K results from the same. Now using content-based filtering on average rating of the movie, we will recommend top N movies to the users (Fig. 8).

```
> recommend_movie(64969)
[1] "Recommendation for movie "
[1] "64969 Yes Man (2008)"
# A tibble: 5 × 5
  movieId                        title                   genres avg_rating no_of_user
*   <int>                        <chr>                    <chr>      <dbl>      <int>
1   51660 Ultimate Gift, The (2006)                   Drama   3.668203        217
2   63179                Tokyo! (2008)                   Drama   3.487179        117
3   68749          Management (2008)        Comedy|Romance   2.887500        160
4   59915                Stuck (2007)        Horror|Thriller   3.072034        118
5   59118 Happy-Go-Lucky (2008)          Comedy|Drama   3.435780        654
> |
```

Fig. 8 Recommendation for "Yes Man"

```
> recommend_movie(58107)
[1] "Recommendation for movie "
[1] "58107 step up 2 the streets (2008)"
# A tibble: 5 × 5
  movieId                                title                                             genres avg_rating no_of_user
*   <int>                                <chr>                                              <chr>      <dbl>      <int>
1   62912 High School Musical 3: Senior Year (2008)                                       Musical   2.423767        446
2   80222                          Step Up 3D (2010)                                 Drama|Romance   3.288934        244
3   53121                    Shrek the Third (2007) Adventure|Animation|Children|Comedy|Fantasy   2.992429       3500
4   98373                  Step Up Revolution (2012)                                       Musical   3.302632        152
5   71537                               Fame (2009)               Comedy|Drama|Musical|Romance   2.656000        125
> |
```

Fig. 9 Recommendation for "Step Up 2"

```
> recommend_movie(52722)
[1] "Recommendation for movie "
[1] "52722 Spider-Man 3 (2007)"
# A tibble: 5 × 5
  movieId                                title                             genres avg_rating no_of_user
*   <int>                                <chr>                              <chr>      <dbl>      <int>
1   53464 Fantastic Four: Rise of the Silver Surfer (2007)    Action|Adventure|Sci-Fi   2.720458       3055
2   95510                  Amazing Spider-Man, The (2012)  Action|Adventure|Sci-Fi|IMAX   3.398940       4245
3   87430                          Green Lantern (2011)     Action|Adventure|Sci-Fi   2.586735       1764
4   57640          Hellboy II: The Golden Army (2008)      Action|Adventure|Fantasy|Sci-Fi   3.373681       3602
5  103042                          Man of Steel (2013) Action|Adventure|Fantasy|Sci-Fi|IMAX   3.221282       3073
> |
```

Fig. 10 Recommendation for "Spider Man 3"

4 Results

The system takes movieId as an input and recommends top five similar movies based on it. Recommended movies rating should be above than 2.4 and at least 100 people should have rated that movie. The complete experiment is performed in R 3.2.2 (Fig. 9).

5 Conclusion

This hybrid model seems to perform good in all the early testing and gives more personalized and accurate results. Genome tag is the key driver for this model along with the content-based filters (Fig. 10).

References

1. Wei S, Zheng X, Chen D, Chen C (2016) A hybrid approach for movie recommendation via tags and ratings. Electron Commer Res Appl 18:83–94
2. Bell RM, Koren Y, Volinsky Chris (2010) All together now: a perspective on the netflix prize. Chance 23(1):24–29
3. Adomavicius G, Tuzhilin A (2005) Toward the next generation of recommender systems: a survey of the state-of-the-art and possible extensions. IEEE Trans Knowl Data Eng 17(6):734–749
4. Linden G, Smith B, York J (2003) Amazon. com recommendations: item-to-item collaborative filtering. IEEE Internet comput 7(1):76–80

5. Sarwar BM, Karypis G, Konstan J, Riedl J (2002) Recommender systems for large-scale e-commerce: scalable neighborhood formation using clustering. In Proceedings of the fifth international conference on computer and information technology, Vol 1
6. Harper FM, Konstan JA (2016) The movielens datasets: history and context. ACM Trans Interact Intell Syst (TiiS) 5(4):19
7. MovieLens Latest datasets. http://files.grouplens.org/datasets/movielens/ml-latest.zip
8. Schafer JHJB, Frankowski D, Herlocker J, Sen S (2007) Collaborative filtering recommender systems. The adaptive web. Springer, Berlin, pp 291–324
9. Herlocker JL, Konstan JA, Borchers A, Riedl J (1999) An algorithmic framework for performing collaborative filtering. In Proceedings of the 22nd annual international ACM SIGIR conference on Research and development in information retrieval, ACM, pp. 230-237
10. Schein AI, Popescul A, Ungar LH, Pennock DM (2002) Methods and metrics for cold-start recommendations. In Proceedings of the 25th annual international ACM SIGIR conference on Research and development in information retrieval, ACM, pp. 253-260
11. Balabanovi M, Shoham Y (1997) Fab: content-based, collaborative recommendation. Commun ACM 40(3):66–72
12. About The Music Genome Project, https://www.pandora.com/about/mgp
13. Vig J, Sen S, Riedl J (2012) The tag genome: encoding community knowledge to support novel interaction. ACM Trans Interact Intell Syst (TiiS) 2(3):13
14. Sarwar B, Karypis G, Konstan J, Riedl J (2000) Application of dimensionality reduction in recommender system-a case study (No. TR-00-043). Minnesota Univ Minneapolis Dept of Computer Science
15. Abdi H, Williams LJ (2010) Principal component analysis. Wiley Interdiscip Rev Comput Stat 2(4):433–459
16. Benesty J, Chen J, Huang Y, Cohen I (2009) Pearson correlation coefficient. Noise reduction in speech processing. Springer, Berlin, pp 1–4
17. Salton G, Wong A, Yang CS (1975) A vector space model for automatic indexing. Commun ACM 18(11):613–620
18. Lee DL, Chuang H, Seamons K (1997) Document ranking and the vector-space model. IEEE softw 14(2):67–75
19. Panigrahi S, Lenka RK, Stitipragyan A (2016) A hybrid distributed collaborative filtering recommender engine using apache spark. Proced Comput Sci 83:1000–1006
20. Lenka RK, Barik RK, Panigrahi S, Panda S (2017) An improved hybrid distributed collaborative model for filtering recommender engine using apache spark. I.J. Intell Syst Appl

Leveraging Machine Learning in Mist Computing Telemonitoring System for Diabetes Prediction

Rabindra Kumar Barik, R. Priyadarshini, Harishchandra Dubey⦿, Vinay Kumar and S. Yadav

Abstract Big data analytics with the help of cloud computing is one of the emerging areas for processing and analytics in healthcare system. Mist computing is one of the paradigms where edge devices assist the fog node to help reduce latency and increase throughput for assisting at the edge of the client. This paper discusses the emergence of mist computing for mining analytics in big data from medical health applications. The present paper proposed and developed mist computing-based framework, i.e., *MistLearn* for application of *K*-means clustering on real-world feature data for detecting diabetes in-home monitoring of patients suffering from diabetes mellitus. We built a prototype using Intel Edison and Raspberry Pi; the embedded microprocessor for *MistLearn*. The proposed architecture has employed machine learning on a deep learning framework for analysis of pathological feature data that can be obtained from smartwatches worn by the patients with diabetes. The results showed that mist computing holds an immense promise for analysis of medical big data especially in telehealth monitoring of patients.

Keywords Diabetes · Cloud computing · Medical big data · Fog · Mist · Edge

R. K. Barik · R. Priyadarshini (✉)
KIIT University, Bhubaneswar, India
e-mail: priyadarshini.rojalina@gmail.com

R. K. Barik
e-mail: rabindra.mnnit@gmail.com

H. Dubey
Center for Robust Speech Systems, The University of Texas at Dallas,
Richardson, TX 75080, USA
e-mail: harishchandra.dubey@utdallas.edu

V. Kumar · S. Yadav
Department of Electronics Engineering, NIT Nagpur, Nagpur, India
e-mail: vk@ece.vnit.ac.in

S. Yadav
e-mail: sadanand.0501@gmail.com

© Springer Nature Singapore Pte Ltd. 2018
M. L. Kolhe et al. (eds.), *Advances in Data and Information Sciences*, Lecture Notes
in Networks and Systems 38, https://doi.org/10.1007/978-981-10-8360-0_9

95

1 Introduction

Cloud computing has the capabilities for sharing and exchange of medical big data belonging to various stakeholders. It has created an environment that enabled wide variety of users to access, retrieve and disseminate medical big data along with associated metadata in a secured manner [1]. Cloud-based framework has leveraged for environmental monitoring, land use and urban planning, natural resource management, marine, coastal, healthcare, and watershed management. There are numerous emerging applications of cloud-based framework. It integrates common medical database operations such as query formation, statistical computations, and overlay analysis with unique visualization functionalities. These features distinguish cloud computing-based framework for decision support systems. It is a widely used tool in public and private sector for explaining events, predicting outcomes, and designing strategies [2].

Medical data contains wide variety of distributions and informative data. In traditional setup of cloud computing-based framework, we send the data to cloud server where these are processed and analyzed [3–5]. This scheme has taken large processing time and required high Internet bandwidth. Fog computing overcomes this problem by providing local computation near the edge of the clients. With the concept of mist computing, it enhances the cloud and fog computing for geospatial data processing by reducing latency at increased throughput. Fog devices such as Intel Edition and Raspberry Pi provide low-power gateway that can increase throughput and reduce latency near the edge of medical-related clients [6]. In addition, it reduces the cloud storage for medical big data. Also, the required transmission power needed to send the data to cloud is reduced as now we send the analysis results to cloud rather than data. This leads to improvement in overall efficiency. Fog devices can act as a gateway between clients such as mobile phones and wearable sensor devices [3]. The increasing use of smart devices led to generation of huge medical big data. Cloud, fog and mist services leverage these data for assisting different analysis. It suggests that the use of machine learning on fog devices which kept close to smart devices for implementing mist computing. This research paper presents a *MistLearn* framework that relied on medical big data analysis on diagnosing patients with diabetes mellitus in India. So, medical big data was processed at the edge and mist using of fog devices and finally has stored at the cloud layer: So, medical data are processed at the edge using of fog devices and finally has stored at the cloud layer. Following contributions are made in the present research paper:

- It gives the detail concepts and architectural framework about the fog and mist computing;
- It presents the big data concept in the field of medical applications with some machine learning techniques in mist computing environments;
- It discusses proposed architecture of *MistLearn* framework that leads to process the various medical data analytics at the edge computing environment;
- There is a case study diagnosing patients with diabetes mellitus which has been elaborated with the use of different mist-assisted fog architecture.

2 Related Works

2.1 Fog Computing

Cloud computing paradigm has the limitation as most of the cloud data centers are geographically centralized. These data centers are not located near the proximity of users or devices [3, 6]. Consequently, latency-sensitive and real-time computation service requests by the distant cloud data centers have often suffered large network congestion, round trip delay, and degraded service quality. Edge computing is an emerging technology for resolving these issues. Edge computing provides the computation nodes near the edge, i.e., close to the clients. It enables the desired data processing at the edge network that consists of edge devices, end devices, and edge server. The smart objects, mobile phones are employed as devices, whereas routers, bridges, wireless access points are employed as edge servers. These components function cooperatively for supporting the capability of edge computation. This paradigm ensures fast response to computational demands and after computation; the analysis report could be transferred to cloud backend. Edge computing is not associated with any kind of cloud services model and communicates more often with end devices. Various combinations of edge and cloud computing lead to the emergence of several computational paradigms such as mobile cloud computing (MCC), mobile edge computing (MEC), and fog computing [7].

Fog computing has been first coined by Cisco in 2012. It is a computing paradigm that decentralizes the resources in data centers toward the different users for improving the Quality of Service (QoS) and experience. This computing does not require computational resources from cloud data centers. In this way, data storage and computation are brought closer to the users leading to reduced latencies as compared to communication overheads with remote cloud servers. It refers to a computing paradigm that uses interface kept close to the devices that acquire data. It introduces the facility of local processing leading to reduction in data size, lower latency, high throughput, and thus power-efficient systems. Fog computing has successfully applied in smart cities and health care [5, 6].

Fog devices are embedded computers such as Intel Edison and Raspberry Pi that act as a gateway between cloud and mobile clients. From the above discussions, we can see that it requires an efficient, reliable, and scalable fog computing-based GIS framework for sharing and analysis of geospatial and medical big data across the Web [8]. Fog computing is a novel idea that helps to reduce latency and increase throughput at the edge of the client with respect to cloud computing environment [3–5]. Fog computing solves the problem by keeping the data closer to local devices and computers, rather than routing through a central data center in the cloud. The simple fog computing architecture is shown in Fig. 1.

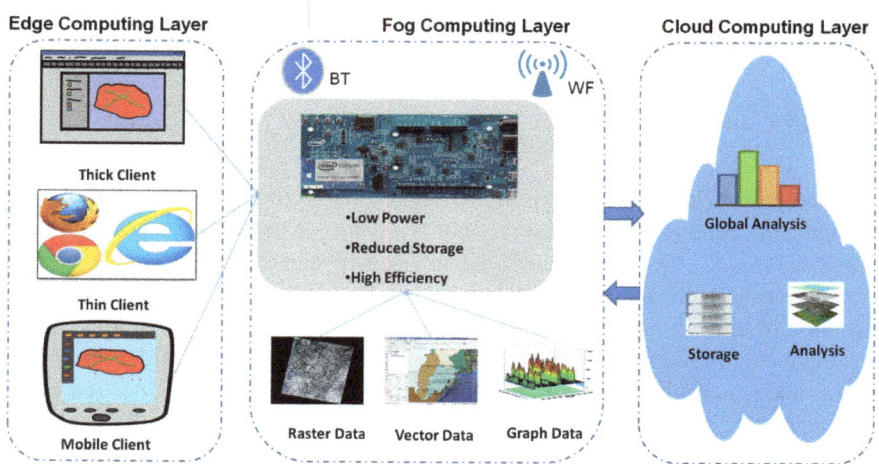

Fig. 1 Fog computing is as an intermediate layer between edge computing layer and cloud layer. The fog computer layer has enhanced the efficiency by providing computing near the edge computing devices. This framework is very much useful for geospatial application, health care, smart city, smart grid and smart home etc. [3]

2.2 Mist Computing

Mist computing has taken edge and fog computing concepts further by pushing some of the computation to the edge of the network, actuator devices, and sensor which build the entire network for cloud data center. With the help of mist computing, the computation has performed at the edge of the network in the microcontrollers of the embedded nodes. The mist computing paradigm decreases latency and increases the autonomy of a solution [9]. Cloud, fog, and mist computing are complementary to each other w.r.t. the application tasks, which are more computationally intensive and can be executed in the gateway of the fog layer while the less computationally intensive tasks can be executed in the edge devices. The processing and the collecting of data are still stored in the cloud data center for the availability to the user. The important application of mist computing is a collection of different services which has been distributed among the computing nodes [10, 11]. Both fog computing and its even younger brother mist computing coined by Cisco and located somewhere between the fog and the edge extend the classical client-server architecture to a more peer-to-peer-based approach, similar or equal to edge [12–15]. By considering this mist computing, the proposed framework, i.e., mist computing, has been sketched for processing of data analysis. Figure 2 has shown the proposed architecture mist computing framework.

The mist computing framework has been categorized into four layers as cloud layer, fog layer, mist layer, and edge layer. In mist computing framework, we used Raspberry Pi in every fog node for better efficiency in time analysis. In this framework, Raspberry Pi Model B Platform has employed. Raspberry Pi consists of a

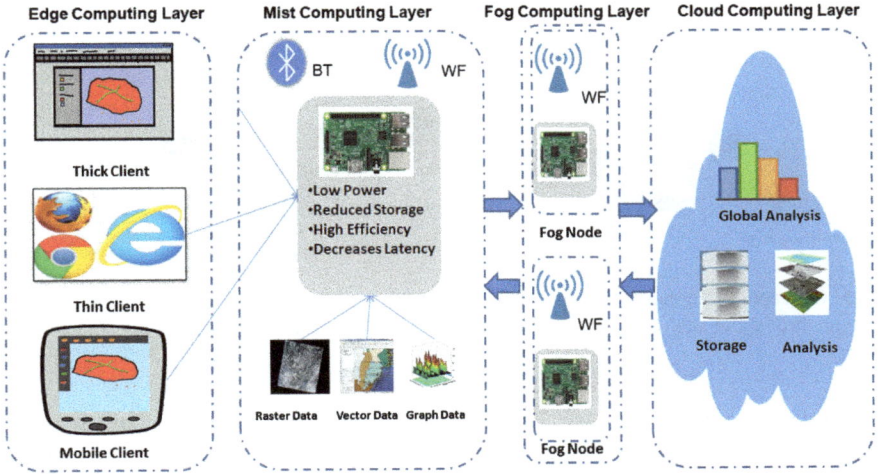

Fig. 2 Conceptual overview of mist computing framework

900 MHz 32-bit quad-core ARM Cortex-A7 CPU with 1 GB RAM. For WiFi connectivity in Raspberry Pi, we used WiFi dongle of Realtek RTL8188CUS chipset.

2.3 Medical Big Data

Big data have distribution, scale, timeliness, and diversity. It requires the analytics to enable insights and the use of new technical architectures that unlock new variety resources of different business value. It includes large datasets whose sizes are beyond the ability of commonly used software tools. It requires special tools to acquire, curate, manage, and process big data in realistic time frames. Big data is generated in multiple forms. Most of the big data is semi-structured, quasi-structured, or unstructured, which makes the processing challenging. Appropriate analysis of big data can discover new correlations to spot business trends, combat crime, and prevent diseases. Big datasets are growing rapidly because they are increasingly gathered by economical and numerous radio-frequency identification (RFID) readers, microphones, aerial (remote sensing), cameras, information-sensing mobile devices, wireless sensor networks, and software logs [3, 5].

Due to diversity of health-related ailments and variety of treatments and outcomes in health sector, there are numerous healthcare data generated that gives rise to the concept of medical big data. Electronic health records, biometric data, clinical registries, medical imaging, patient-reported data, and different administrative claim record are the main sources for medical big data. Medical big data have several typical features that are different from big data from other disciplines. These big data are often hard to access, and investigators in the medical area are aware of risks related

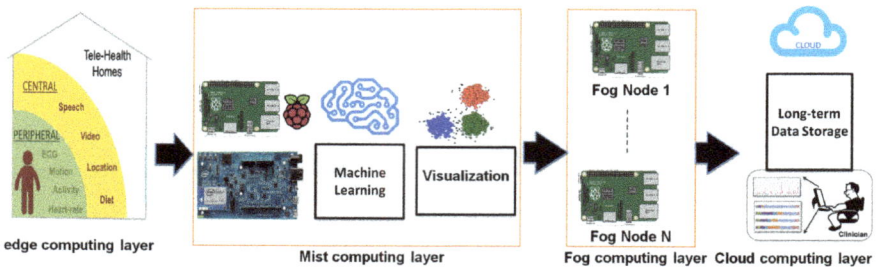

Fig. 3 Conceptual diagram of the proposed *MistLearn* framework for power-efficient, low latency, and high throughput analysis of medical big data between edge and fog layer

to data privacy and access [16]. Big data poses some challenges for data analysis with cloud computing. Reliability, manageability, and low cost are the key factors that make cloud computing useful for data processing [17, 18]. However, the security and privacy are the main concerns for processing sensitive data. Particularly in health geo-informatics applications with sensitive data, we require secure processing. For minimizing privacy and security threats, the data should be utilized as per the user's context for limited amount of data access within the model. After processing, the data should be transferred to the next level for final processing of data analysis, that will benefit the data privacy and security [3, 5, 16].

3 Proposed Model

3.1 *Proposed Mist Architecture:* **MistLearn**

This section describes various components of the proposed architecture, i.e., *MistLearn,* and discusses the hardware, software, and methods implemented for machine learning approaches on medical big data. We employed Intel Edison and Raspberry Pi as mist computing device in proposed framework. Intel Edison is powered by a rechargeable lithium battery and contains dual-threaded 500 MHz, dual-core and Intel Atom CPU along with a 100 MHz Intel Quark microcontroller. It possesses 1 GB memory with 4 GB flash storage and supports IEEE 802.11 a/b/g/n standards. It connects to WiFi and based on UbiLinux operating system for running compression utilities. Figure 3 has shown the proposed architecture which employs mist gateway between edge and fog layer.

The proposed *MistLearn* framework has three layers as client-tier layer, fog computing, and cloud computing layer. In client-tier layer, the categories of users divided into mobile client, thin client, and thick client environments. Processing of medical big data can be possible within these three environments. Cloud computing layer is mainly focused on overall storage and analysis of medical big data. The mist computing layer worked as middle-tier between client-tier layer and fog layer. We experimentally validated the mist computing layer as low-power consumption, overlay analysis capabilities, and reduced storage requirement. In mist layer, all mist nodes are developed with Intel Edison processor for processing of medical data.

MistLearn framework used to assist and hence enhanced the capabilities of fog and cloud computing framework. The mist layer in *MistLearn* processes the data, and after processing, it has the ability to send the data to cloud layer for long-term analysis and storage. So, this framework enables the more power to the end users for better performance without computational overhead at cloud layer. The designed framework has the benefit when computational overhead at cloud layer has very high and data sizes are expanding. Thus, the mist layer increases the overall efficiency by reducing the latency and increasing the throughput. This framework added privacy benefit where we processed the data locally at mist devices and sent only the analysis results to fog and then to cloud computing layer. In the proposed framework, the number of mist devices has been utilized according to application demand. For example, in case we have large amount of data that has to be processed in short time, we need more mist and fog nodes.

4 Case Study: Prediction of Diabetes in Telemonitoring System

4.1 Diagnosing Patients with Diabetes Mellitus

Diagnosing the diabetes mellitus is highly required in telehealth monitoring system. The way of treatment required for a diabetic person is different than a normal person. For detecting whether a person is diabetic or not, the features which are taken in this work are (1) plasma glucose concentration (2) diastolic blood pressure (3) 2-h serum insulin (mu U/ml) (4) body mass index (5) diabetes pedigree function (6) age in years. These data can be collected using health sensors embedded smartphones. For experimental purpose, a standard dataset from UCI learning repository (C.L. Blake, C.J. Merz) is taken and used. In this dataset, there are 768 samples present, out of which 600 samples are employed as training and 168 samples are employed for testing purpose [19, 20].

The proposed model is based on handling medical big data. So, for accomplishing the task of classification, conventional classification tools like neural networks are substituted by deep neural network. The classification of diabetes and non-diabetes people is done by using a simple deep neural network. A deep neural network model is configured using Deeplearning4j tool. Deeplearning4j is Java-based open-source library for creating and deploying deep neural network. These libraries are customized through an IntelliJ idea which is an open-source Java IDE [21]. The model is tailored according to the suitability of the current data. The classification accuracy is coming as 81.89%. This measure is taken on an averaged 10 random iterations. Figures 4 and 5 have shown the plotting of classification result of training and test datasets.

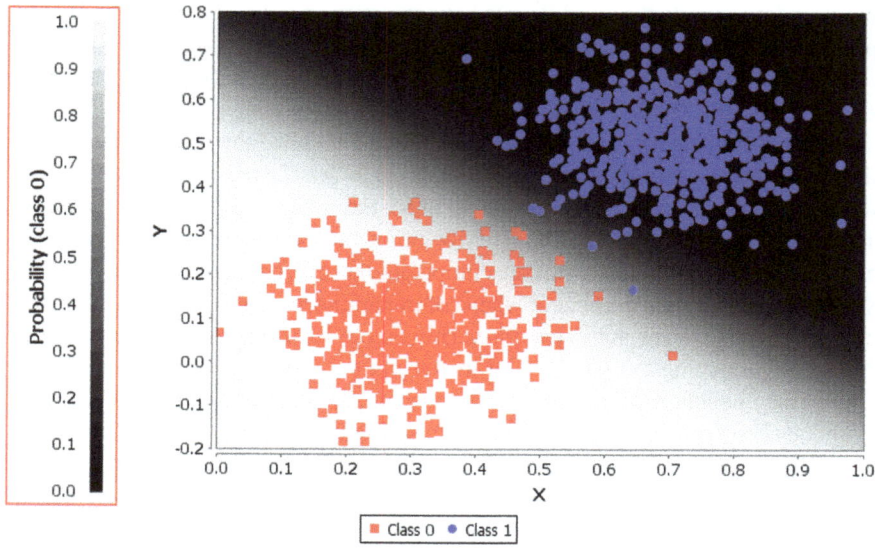

Fig. 4 *K*-means clustering plot for two clusters on training

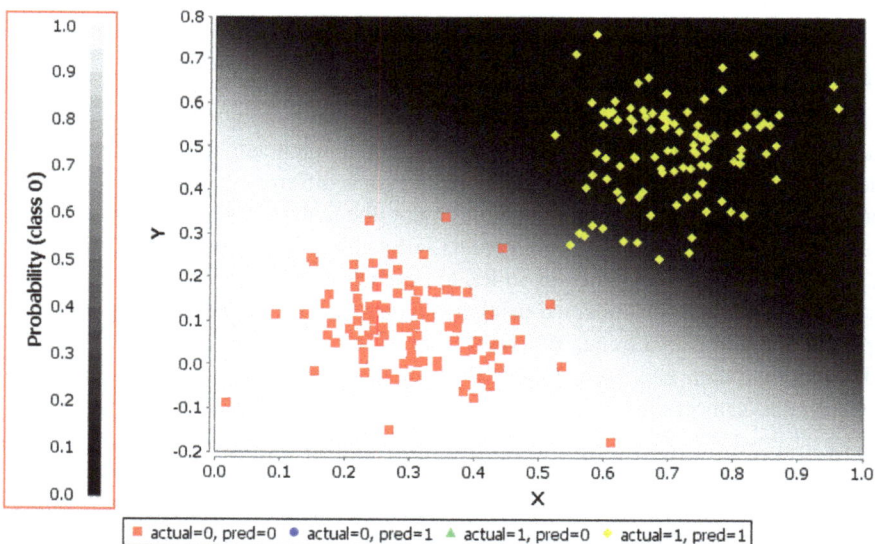

Fig. 5 *K*-means cluster plot for test dataset

5 Concluding Remarks

In this paper, diagnosing the diabetes mellitus data from in-home monitoring of patients with diabetes disease was considered for case study using *MistLearn* architecture. We further leveraged machine learning approaches such as deep neural network on well-type data from diabetes data. In future, we would like to add more intelligent processing procedures. We also validated *MistLearn* architectures for application-specific case studies. Intel Edison processor and Raspberry Pi were used as mist and fog processors in mist/fog computing layers. Mist nodes not only reduce storage requirements but also result in efficient transmission at improved throughput and latency. Mist of things is collection of all nodes between client layer and cloud via fog. The edge computing done on mist/fog nodes creates an assistive layer in scalable cloud computing. With increasing use of wearable and Internet-connected sensors, enormous amount of data is being generated.

The cloud could be reserved for long-term analysis. Mist computing emphasizes proximity to end users unlike cloud computing along with local resource pooling, reduction in latency, better user experiences, and better Quality of Service. This paper relied on mist node for low-resource machine learning. As a use case, we employed deep neural network to classify and predict diabetes disease on patients with diabetes. Proposed *MistLearn* architecture is showing around 81% of accuracy in prediction of analysis. Mist computing has reduced the onus of dependence on cloud services with availability of big data. There will be more aspects of this proposed architecture that can be investigated in future. We can expect mist architecture to be crucial in shaping the way big data handling and processing happens in near future.

References

1. Chen Z, Chen N, Yang C, Di L (2012) Cloud computing enabled web processing service for earth observation data processing. IEEE J Sel Top Appl Earth Obs Remote Sensing 5(6):1637–1649
2. Huang Q, Yang C, Liu K, Xia J, Xu C, Li J, Gui Z, Sun M, Li Z (2013) Evaluating open-source cloud computing solutions for geosciences, Comput Geosci 5941–5952
3. Barik RK, Dubey H, Samaddar AB, Gupta RD, Ray PK (2016) FogGIS: fog computing for geospatial big data analytics. In: IEEE Uttar Pradesh section international conference on electrical, computer & electronics engineering, pp 613–618
4. Patra SS, Barik RK (2015) Dynamic dedicated server allocation for service oriented multi-agent data intensive architecture in biomedical and geospatial cloud. In: Cloud technology: concepts, methodologies, tools, and applications. IGI Global, pp 2262–2273
5. Dubey H, Yang J, Constant N, Amiri AM, Yang Q, Makodiya K (2015) Fog data: enhancing telehealth big data through fog computing. In: ACM proceedings of the ASE BigData & social informatics, p 14
6. Dubey H, Constant N, Monteiro A, Abtahi M, Borthakur D, Mahler L, Sun Y, Yang Q, Mankodiya K (2017) Fog computing in medical internet-of-things: architecture, implementation, and applications. Handbook of large-scale distributed computing in smart health-care. Springer International Publishing AG, Cham

7. Chiang M, Zhang T (2016) Fog and IoT: an overview of research opportunities. IEEE Internet Things J 3(6):854–864
8. Yang C, Michael G, Qunying H, Doug N, Robert R, Yan X, Myra B, Daniel F (2011) Spatial cloud computing: how can the geospatial sciences use and help shape cloud computing? Int J Dig Earth 4(4):305–329
9. http://www.thinnect.com/mist-computing/. Accessed on 17 July 2017
10. http://rethinkresearch.biz/articles/cisco-pushes-iot-analytics-extreme-edge-mist-computing-2/. Accessed on 17 July 2017
11. Orsini G, Bade D, Lamersdorf W (2015) Computing at the mobile edge: designing elastic android applications for computation offloading. In: IEEE IFIP wireless and mobile networking conference (WMNC), pp 112–119
12. Preden JS, Tammemäe K, Jantsch A, Leier M, Riid A, Calis E (2015) The benefits of self-awareness and attention in fog and mist computing. Computer 48(7):37–45
13. Uehara M (2017) Mist computing: linking cloudlet to fogs. In: International conference on computational science/intelligence & applied informatics. Springer, Cham, pp 201–213
14. Barik RK, Tripathi A, Harishchandra D, Lenka RK, Pratik T, Sharma S, Mankodiya K (2017) MistGIS: optimizing geospatial data analysis using mist computing. In: International Conference on Computing Analytics and Networking
15. Barik RK, Priyadarshini R, Dubey H, Kumar V, Mankodiya K (2018) FogLearn: leveraging fog-based machine learning for smart system big data analytics. Int J Fog Comput vol 1(1). ISSN: 2572-4908
16. Lee CH, Yoon HJ (2017) Medical big data: promise and challenges. Kidney Res Clin Pract 36(1):3
17. Gupta A, Merchant PS (2016) Automated lane detection by K-means clustering: a machine learning approach. Electron Imaging 14:1–6
18. Lenka RK, Barik RK, Mishra A, Sharma S, Dubey H, Simha NVR, Mankodiya K (2017) GeoTour: leveraging cloud GIS model for tourism information infrastructure network. In: 2nd international conference on computer, communication and computational sciences
19. Priyadarshini R, Barik RK, Dash N, Mishra BKK, Misra R (2017) A hybrid GSA-K-mean classifier algorithm to predict diabetes mellitus. Int J Appl Metaheuristic Comput (IJAMC) 8(4):99–112
20. Internet-1 CL, Blake CJ, Merz (2017) UCI repository of machine learning databases. Available from: http://www.ics.uci.edu/-mlearn/MLRepository.html. Accessed on 6 Feb 2017
21. Internet-2: Skymind (2016) Deeplearning4j. Available: http://deeplearning4j.org/. Accessed on 6 Feb 2017

A Novel Energy-Efficient Hybrid Full Adder Circuit

Trapti Sharma and Laxmi Kumre

Abstract This article presents a novel energy-efficient hybrid one-bit full adder cell employing modified GDI logic and transmission gate logic. To analyze the performance of various circuits, simulations were carried out using Synopsys HSPICE tool taking 45-nm technology model. Full-swing differential XOR–XNOR circuits with restoration transistors are employed in order to realize a noise-resistant full adder circuit. The weak restoration transistors at the intermediate outputs of XOR–XNOR ensure the reduced static power dissipation together with the usage of low power transmission gates at the output side leads to overall reduction in power of the proposed circuit. Comprehensive experiment results illustrate the superiority of the proposed adder in terms of power–delay product (PDP) over the other existing designs with regard to the different simulation conditions such as supply power scaling, load, temperature, and frequency variations.

Keywords Hybrid full adder · Low power · Energy efficient · VLSI design

1 Introduction

With the emergence of portable electronic devices such as mobile phones, laptops, notebooks, personal communication system, personal digital assistants etc., has intensified the requirement of energy efficient system. together with the shrinking technology evolution in electronic industry. Many attempts have been taken by the researchers to frame highly efficient ultra large-scale integration designs with high throughput. The critical performance metrics accounted for an energy-efficient digital systems in microelectronics are power consumption, processing speed, and area overhead. Power consumption is the primary parameter considered in the digital system design. As the power dissipation of the system rises, the extra accessories to make the system resistant against heat will be required which in turn augments the cost, weight, size, and overall complexity of a digital system. On the other hand

T. Sharma (✉) · L. Kumre
Department of ECE, Maulana Azad National University of Technology, Bhopal, India
e-mail: trapti16sharma@gmail.com

in high-speed electronic era, the operating speed is also the primal requirement. To attain high processing speed, high clock rates could be employed but it will make vulnerable impact on overall power dissipation of the circuit. Therefore to accomplish a compromise between the contradictory attributes, a new quantitative metric is evaluated termed as a power–delay product (PDP). It is widely used figure of merit for the performance assessment of the digital circuits.

Several VLSI systems such as microprocessor and microcontroller system, application-specific digital signal processors (DSPs), image and video processors [1] extensively use basic arithmetic operations such as addition, subtraction, multiplication and multiply and accumulate [2]. Addition is the basic function which in turn could be utilized to realize all other remaining operations. Therefore, one-bit full adder cell lies in the critical path of the most of the digital systems [3]. The overall performance of the entire complex systems solely depends on the performance of the nucleus adder circuit. Hence, the enhancement in the performance of the core adder circuit results in the overall performance enhancement at system level [4]. Several static [5] and dynamic logic style [6] full adder circuits are reported in the literature. The static logic style is most commonly used for various ultra low-power digital circuit implementation as they have more simplicity, reliability, and less power requirement as compared to dynamic logic style. Also, dynamic logic style has shortcomings such as charge sharing, less immunity toward noise, high clock load. Therefore, all the static designs are included for discussion in this article.

Most of the static full adder circuits available in the literature employ three logic styles namely CMOS logic, pass transistor logic (PTL), and transmission gate logic. Each of the logics has its pros and cons. The classical complementary metal–oxide–semiconductor logic (CMOS) full adder circuit is presented in [7]. CMOS logic is based on dual networks of PMOS and NMOS to implement a logic function. It provides the benefits of robustness against transistor sizing and voltage scaling, low static dissipation, and full-swing output signals. But circuit becomes more complex as it requires a large number of transistors which result in increased circuit capacitances, increased power consumption, and long propagation delays. To overcome the shortcomings of CMOS logic leads to the development of new logic design suites. Another popular logic emerge after that is PTL in which source side is fed with the inputs rather than the voltage source. Some of the advantages of the PTL logic are low power dissipation and small area due to reduction in the number of the transistors, high-speed and lower interconnection effects [8]. However it suffers from slower operation at reduced supply voltages, high static power dissipation and threshold voltage drop problem which produces non-full-swing outputs. Some of the PTL adder circuits are reported in [4, 6, 9, 10]. Another extended version of PTL logic is gate diffusion input logic (GDI) whose structure resembles like a CMOS inverter. GDI facilitates to design digital circuits with small area and improved power characteristics due to significant reduction in subthreshold and gate leakage component [9]. Also, some of the basic functions such as AND, OR, multiplexer could be realized with only two transistors [11]. However, GDI circuits suffer from threshold voltage drop problem, which results in significant degradation at output side [12]. To resolve the issue of low logic swing in GDI logic, swing restoration buffers [13] are connected at the output; however, it results in increase in transistor count and power dissipation. Another approach is to employ multiple threshold voltage technique [9] as there is a flexibility to operate at different threshold voltages in nanoscale region. Low threshold transistors are used, where low voltage drop is desired and high threshold transistors are used for inverters. Some of the GDI full adder circuit designs are presented in [6, 11, 14]. The full adder circuit using transmission gate is reported in [7]. It resolves the problem of low logic level swing encountered in PTL and GDI logic, but again the compromise with the area has to be done. Another alternative to design a full adder circuit which uses more than one logic style in order to achieve the ultimate goal of overall performance enhancement known as hybrid logic styles. Hybrid adders are discussed in [10, 13, 15–17]. In this paper, a new full-swing differential design of XOR–XNOR gate is proposed and by using that module an ultra low-power hybrid full adder cell is designed and simulated at 45-nm technology using HSPICE. The main objective of this article is to improve various performance parameters of the full adder circuit as compared to already existing ones. The paper is organized as follows: Sect. 2 briefly reviewed the previously published full adder circuits. In Sect. 3, novel full adder circuit is discussed. Simulation results and performance analysis are shown in Sect. 4. Finally in Sect. 5, conclusion is made.

2   Previous Works

2.1   The SRCPL Full Adder

The structure of swing-restored complementary pass transistor logic [SRCPL] full adder shown in Fig. 1a is reported in [12]. It uses pass transistor powerless/groundless logic styles to realize a full adder cell. A multiplexer with A XOR B and A XNOR B as the inputs and C as a selection input signal is used to generate the SUM output. Another multiplexer with $A \cdot B$ and $A + B$ as inputs and C as a selection input is used to produce the carry signal. Although this design has a large number of transistors, it has lower power consumption due to very low static consumption and non-availability of direct rail paths between the inputs and the outputs. Transmission gates are employed at the last stage of the circuit to generate the output signals which result in full-swing output signals. Also, the critical path between the input carry signal and output carry signal consists of three transistors which makes the circuit to operate at high speed. The main impediment seen in the SRCPL is that it lacks sufficient driving capability while working in cascaded mode. Its performance is drastically affected when encapsulated in wide cascaded configuration.

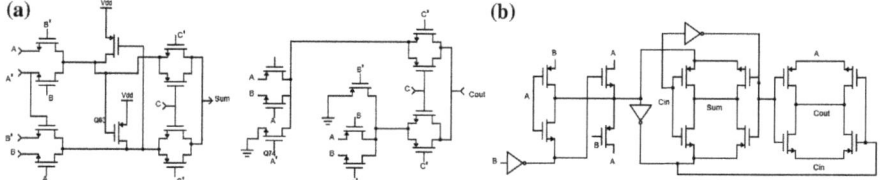

Fig. 1 **a** SRCPL full adder [12], **b** FSGDI adder [11]

2.2 The FSGDI Design

The full-swing gate diffusion input adder circuit [FSGDI] with 18 transistors shown in Fig. 1b is reported in [11] with the goal to achieve a faster cascaded operation. The author claims to have lowest energy consumption and resistant against various process variations. Full-swing XOR gates using GDI logic are used to design the full adder cell. To remove the threshold voltage loss problem in XOR gate, two level restoration transistors are inserted to recover the weak 0 or 1 output respectively. The multiplexers using GDI logic are used to realize the output signals with XOR and XNOR as the intermediate inputs. Its critical path consists of an inverter which results in increment of the delay of the overall circuit.

2.3 SERF Full Adder

SERF full adder [18] shown in Fig. 2a is amalgam of improved form of sense energy recovery full (SERF) adder design [19] and GDI logic. In SERF logic, the capacitance charged during high logic is consumed to make the gate logic high instead of draining it to ground, which results in less energy consumption and low power design. Two cascaded XNOR cells gives the SUM output and carry is implemented with GDI multiplexer. As compared with previous GDI circuits, this design has no threshold voltage drop problem without inserting any level restoration buffers. The main problem encountered in SERF cells is that supply voltage scaling could not be reduced below 2Vtn + Vtp. Otherwise, it performs satisfactorily well at high supply voltages. Also, it has long critical path delay which causes the low-speed operation, especially at low voltages.

2.4 HFAD Design

A hybrid full adder design (HFAD) shown in Fig. 2b has been reported in [15]. The circuit is implemented by the combination of complementary metal–oxide–semiconductor (CMOS) logic and transmission gate logic. This circuit contains the

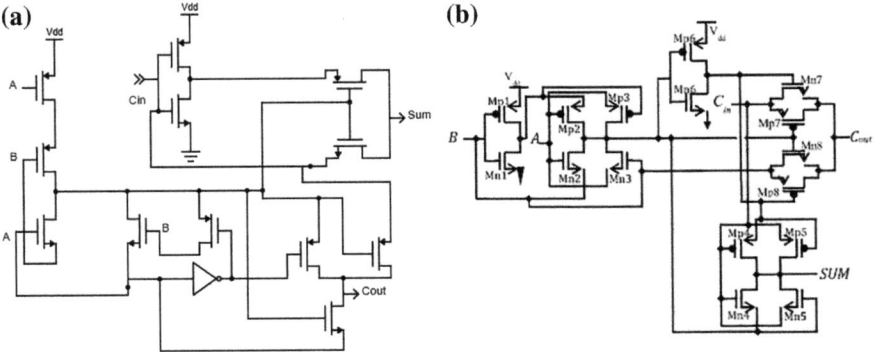

Fig. 2 **a** SERF adder [18], **b** HFAD adder [15]

modified XNOR circuit of six transistors which provides the low power consumption and high-speed operation. The decrement in the power consumption is made by optimized sizing of the transistors in inverter circuit. The channel width of the transistors Mn7, Mp7, Mn8, and Mp8 is made large which results in the overall reduction of the carry propagation delay. Level restoring transistors Mp3 and Mn3 avail the full-swing output signals. Modified XNOR circuit with level restoration transistors is used to generate the sum output while XNOR–XOR input based transmission gates constitute the carry generation unit. However, one of the essential drawbacks of this circuit is that when it operates in carry propagation mode, high-speed operation is hampered and its performance degrades further as the number of cascading adder stages increases. The buffers were included in between the cascaded stages at the appropriate stage in order to minimize the overall delay.

3 Proposed Full Adder

In this section, the proposed hybrid full adder circuit is based on modified GDI logic and the transmission gate logic. Proposed adder circuit shown in Fig. 5 satisfies the following equation:

$$SUM = (B \oplus C).A + (B\bar{\oplus}C).\bar{A} \tag{1}$$

$$Cout = (B \oplus C).B + (B\bar{\oplus}C).A \tag{2}$$

Proposed circuit contains mainly two modules namely differential XOR–XNOR and transmission gate-based 2:1 multiplexer circuit. In this article, a novel full-swing differential XOR–XNOR circuit is proposed and embedded with multiplexer to form a complete adder circuit. As seen from the above equation, the basic unit comprising the whole module basically consists of XOR–XNOR module. A new design for differential XOR–XNOR circuit using transistors M1, M2, M3, and M4

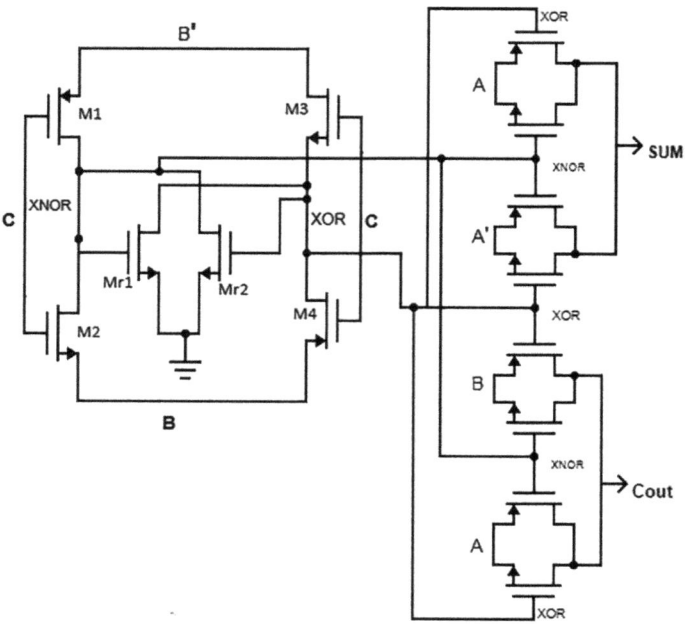

Fig. 3 Proposed circuit

is shown in Fig. 3. Differential XOR–XNOR circuit uses only four transistors with two restoration transistors Mr1 and Mr2. It utilizes the modified GDI logic [14] for the logic implementation. In modified GDI logic [14], the NMOS bulk terminal is connected to the ground and the bulk terminal of PMOS is tied to Vdd which also helps in marginal swing improvement at the output side. Table 1 elucidates the operation of proposed XOR–XNOR cell. The output of the differential module suffers from the threshold voltage drop problem. As seen from Table 1 when the input vector BC = 00, then XOR gets a degraded zero output, and for the input vector BC = 10, at XNOR output, the weak low signal is observed. This threshold loss is not desirable as it negatively affects the power–delay characteristics of the overall circuit. Therefore, it is necessary to sort out the above issue before that submodule is embedded in larger complex module. To rectify this threshold voltage drop problem in PTL circuits, feedback loop method across the output side is adopted as explained in [20]. Mr1 and Mr2 are the feedback loop transistors that restore the degraded levels of the output. Feedback loop transistors decrease the power consumption as well as delay by providing the full-swing outputs.

Table 1 Operation of differential XOR–XNOR

B	C	XOR	XNOR
0	0	poor 0	strong 1
0	1	strong 1	strong 0
1	0	strong 1	poor 0
1	1	strong 0	strong 1

4 Simulation Results and Comparison

In this section, comprehensive computer simulations are performed to evaluate the performance of the various full adder circuits discussed in the above sections. Simulations of different adder cells are carried out using Synopsys HSPICE tools for transistor with 45-nm channel length. In order to evaluate the performance, the simulations are performed at various supply voltages, temperature, frequencies, and loads. The proposed design is compared with other state-of-the-art designs cited in the literature [11, 12, 15, 18]. The sizing of the transistors plays a major role in minimizing the power consumption of the proposed circuit. In the proposed design, weak inverters with width same as channel length are used. On the other hand, strong transmission gates are realized with large width (W/L = 2:1). Level restoration transistors are made with smaller width taken half of channel length. In order to make fair comparison, the transistor sizes for all the remaining circuits were taken as taken by the correspondent author. The propagation delay is measured from the instant the input signal reaches 50% of the input supply voltage until the output signal attains the same level. The corresponding delay for both the outputs SUM and Cout is calculated and then greater among them is reported as final delay of the circuit. Furthermore, power consumption taken is the average power measured for the given interval. To achieve a trade-off between power consumption and delay, an optimized parameter power–delay product (PDP) is considered and minimized for the proposed circuit. In the first experiment, the circuits are tested at different supply voltages namely 0.8, 1, and 1.2 V all at 100 MHz frequency with 1fF load capacitance. The detailed results of this simulation are exhibited in Table 2. There is approximately 85, 90, and 10% improvement in power consumption of the proposed circuit as compared with FSGDI, SERF, and HFAD respectively. Also, the proposed design is faster of about 90, 74, and 43% as compared with SRCPL, FSGDI, and SERF circuits. And finally, improvement in the PDP of about 90, 96, 94 and 10% is seen as compared with SRCPL, FSGDI, SERF, and HFAD, respectively. PDP vs supply voltage variation graph is plotted as shown in Fig. 4a. From the graph, it is illustrated that at all the supply voltages taken, the PDP of proposed design is at its minimum level.

To examine the driving capability of the circuit, they are simulated against variety of output load capacitors ranging from 1 up to 4.2 fF at 100 MHz frequency and 1 V supply voltage. The PDP of different designs, against load capacitor variation, is shown in Fig. 4b. From the experimental results, it could be observed that PDP

Table 2 Performance metrics against power supply variation at freq = 100 MHz and C = 1 ff

Vdd(V)	0.8	1	1.2
Delay (ps)			
SRCPL [12]	354.80	152.2	96.108
FSGDI [11]	39.380	60.092	39.836
SERF [18]	39.894	26.938	14.166
HFAD [15]	21.588	15.351	10.590
Proposed	25.972	15.315	10.138
Power (uW)			
SRCPL [12]	0.092096	0.13206	0.18113
FSGDI [11]	0.51633	0.84116	0.12842
SERF [18]	0.14735	1.3219	1.1820
HFAD [15]	0.14530	0.14355	0.16263
Proposed	0.11462	0.12955	0.14783
PDP (E-18)			
SRCPL [12]	32.676	20.104	17.408
FSGDI [11]	77.540	50.547	51.157
SERF [18]	58.783	35.609	16.744
HFAD [15]	3.1368	2.1984	1.7222
Proposed	2.977	1.984	1.4986

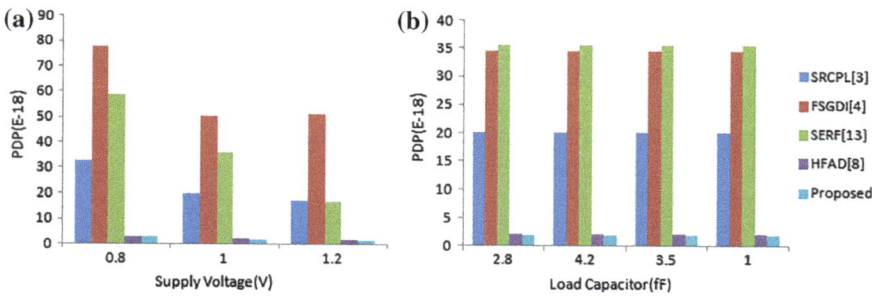

Fig. 4 **a** PDP versus supply voltage, **b** PDP versus load capacitor

(1.984e-18) of the proposed circuit is lowest in comparison with all other designs considered for all the load capacitor values taken. Another important trait taken into account is the circuits immunity toward the ambient temperature variations. In order to estimate the susceptibility to the temperature noise, the circuit is simulated in a vast range of temperature from 0 to 70 °C at 1 V input voltage, 100 MHz frequency, and 1fF load capacitance. The results of this experiment are plotted as shown in Fig. 5a. It can be inferred from the experimental results that proposed circuit has robust operation and admissible functionality across the wide range of temperature. Also, it outperforms the other designs in terms of PDP. To estimate the performance

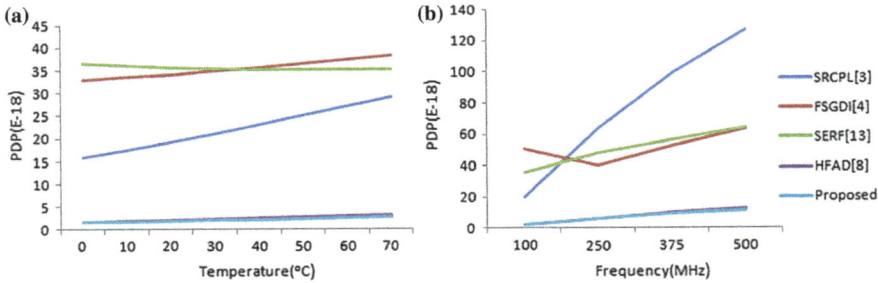

Fig. 5 **a** PDP versus temperature, **b** PDP versus input operating frequency

of circuits at different frequencies, they are enforced at 100–500 MHz frequency, 1 V supply voltage, and 1 fF load capacitance. The result of this experiment is depicted in Fig. 5b. According to the resultant data, it is implied that proposed circuit has lowest PDP as compared to other circuits and works reliably at different frequencies.

5   Conclusion

In this article, a hybrid full adder circuit is designed using the synergic combination of modified GDI logic and transmission gate logic. In order to estimate the performance of various adder circuits, simulations were carried out using Synopsys HSPICE tool at 45-nm technology. Investigation at various supply voltages, load, temperature, and frequencies were considered. This design adopts the full-swing differential XOR–XNOR circuit to mitigate the threshold loss problem. The goal of low power is achieved through the coupling of strong transmission gates, weak inverters, and level restoring transistors. Results of the simulation indicate that the proposed adder circuit has considerable improvements in terms of power and power–delay product (PDP) in comparison with other previously existing designs. The savings in PDP sweeps from 96 to 10%, and in terms of power, it is 85% to 10%, respectively. Hence, the achievement of lowest PDP under various process conditions favors the suitability of the proposed circuit for energy-efficient arithmetic applications.

Acknowledgements This work has been done from the Grant Received from Visvesvaraya PhD Scheme for Electronics and IT.

References

1. Sabaghi M, Marjani S, Majdabadi A (2016) The design of ultra-low power adder cell in 90 and 180 nm CMOS technology. Circuits Syst 7(2):58–67

2. Wariya S, Nagaria R, Tiwari S (2012) Performance analysis of high speed hybrid CMOS full adder circuits for low voltage VLSI design. VLSI Design 1–18
3. Jiang Y, Al-Sheraidah A, Wang Y, Sha E, Chung J-G (2004) A novel multiplexer-based low-power full adder. IEEE Trans Circuits Syst II 51(7):345–348
4. Wang D, Yang M, Cheng W, Guan X, Zhu Z, Yang Y (2009) Novel low power full adder cells in 180 nm CMOS technology. ICIEA 1:0
5. Zimmermann R, Fichtner W (1997) Low-power logic styles. IEEE J Solid State Circuit 32:1079–1090
6. Pashaki ER, Shalchian M (2016) Design and simulation of an ultra-low power high performance CMOS logic: DMTGDI. Integr VLSI J 55:194–201
7. Weste NHE, Eshraghian K (1988) Principles of CMOS VLSI design: a system perspective. Addison-Wesley, Reading
8. Morgenshtein A, Fish A, Wagner IA (2002) Gate diffusion input (GDI)-A power efficient method for digital combinatorial circuits. IEEE Trans VLSI Syst 10(5):566–581
9. Morgenshtein A, Shwartz I, Fish A (2010) Gate diffusion input (GDI) logic in standard CMOS nanoscale process. In: Proceedings of the IEEE 26-th convention of electrical and electronics engineers in Israel
10. Chowdhury SR, Banerjee A, Roy A, Saha H (2008) A high speed 8 transistor full adder design using novel 3 transistor XOR gates. Int J Electron Circuits Syst II:217–223
11. Shoba M, Nakkeeran R (2016) GDI based full adders for energy efficient arithmetic applications. Eng Sci Technol Int J 19:485–496
12. Aguirre-Hernandez M, Linares-Aranda M (2011) CMOS full-adders for energy-efficient arithmetic applications. IEEE Trans Very Larg Scale Integr (VLSI) Syst 19(4):718–721
13. Foroutan V, Taheri M, Navi K, Mazreah A (2014) Design of two low power full adder cells using GDI structure and hybrid CMOS logic style. Integration (Amst) 47(1):48–61
14. Uma R, Dhavachelvan P (2012) Modified gate diffusion input technique: a new technique for enhancing performance in full adder circuits. In: Proceedings of ICCCS, pp. 74–81
15. Bhattacharyya P, Kundu B, Ghosh S, Kumar V, Dandapat A (2015) Performance analysis of a low-power high-speed hybrid 1-bit full adder circuit. IEEE Trans Very Larg Scale Integr (VLSI) Syst 23(10):2001–2008 Oct
16. Goel S, Kumar A, Bayoumi M (2006) Design of robust, energy-efficient full adders for deep-submicrometer design using hybrid-CMOS logic style. IEEE Trans Very Large Scale Integr (VLSI) Syst 14(12):1309–1321
17. Belgacem H, Chiraz K, Rached T (2012) A novel differential XOR-based self-checking adder. Int J Electron 99(9):1239–1261
18. F. Moradi, D.T. Wisland, H. Mahmoodi, S. Aunet, T.V. Cao, A. Peiravi, Ultra low power full adder topologies, in: Proceedings of the IEEE International Symposium on Circuits and Systems, pp. 3158–3161, May 2009
19. Shalem R, John E, John L.K (2002) A novel low power energy recovery full adder cell. In: Proceedings IEEE great lakes VLSI Symposium, pp. 380–383
20. Morgenshtein A, Shwartz I, Fish A (2014) Full swing gate diffusion input (GDI) logic case study for low power CLA adder design. Integr VLSI J 47(1):62–70

An Insight into Theory-Guided Climate Data Science—A Literature Review

Rafiya Sheikh and Sunita Jahirabadkar

Abstract Data science models, though successful in a large number of commercial domains, have found limited applications in scientific problems that involve complex physical phenomena. Most of these problems comprise of multi-spectral data composites. Climate science and hydrology is one such scientific domain that faces several big data challenges. Climate data poses many challenges in research because of its spatiotemporal characteristics, high degree of variance, and predominantly its physical nature. One such challenging data in climate science and hydrology is precipitation data. Precipitation data is vast, and generated at a fast pace from several sources, but due to the lack of underlying principles, the models in data science to address climatic issues such as precipitation are dysfunctional. These challenges call for a novel approach that integrates domain knowledge and data science models. To do so, the paper surveys an evolving paradigm of theory-guided data science (TGDS). It is a new paradigm in data science and analytics that aims to improve the generalization of data science models and improve their effectiveness in scientific discovery. The authors, through the survey, present the challenges imposed by climate data, which is representative of the precipitation data, and limitations of traditional data science methods. The paper suggests a shift in data science practices to adapt theory-guided data science for climate and hydrology domain of precipitation data, by providing insights on TGDS, its models and approaches.

Keywords Data science · Theory-guided · Knowledge discovery · Climate change · Climate science · Precipitation

R. Sheikh (✉) · S. Jahirabadkar
Department of Computer Engineering, M.K.S.S.S. Cummins College
of Engineering for Women, Pune, India
e-mail: rafiya.sheikh@cumminscollege.in

S. Jahirabadkar
e-mail: sunita.jahirabadkar@cumminscollege.in

© Springer Nature Singapore Pte Ltd. 2018
M. L. Kolhe et al. (eds.), *Advances in Data and Information Sciences*, Lecture Notes
in Networks and Systems 38, https://doi.org/10.1007/978-981-10-8360-0_11

115

1 Introduction

Climate change has been an international concern; from the suspicion of the effect of greenhouse gases to global warming, climate change has been debated across countries and scientific experts for decades now. The research in the climate science is inherently interdisciplinary. The research started with numerical models and historical data storage which have led to an improved understanding of the state of climate affairs by helping understand the correlation between signs from vital climate events. The observational and historic data for climate change is big data and is ever increasing and ubiquitous. In the fast-paced technology era, the data deluge prevails in almost every sector, from medicine to social network to digital cash transactions with credit card data mounding up to the electronic health records adding to this deluge [1–3]. Climate data for precipitation too is a part of the data deluge. Climate data, such as precipitation data, is global in nature, very vast, very erratic, ever changing, unpredictable, and further includes many such facets to it. This poses immense challenges to data scientists as they cannot rely on traditional data science models to a good extent. Climate data can serve as an effective tool in making lives easier on the planet by unprecedented edgy predictions. Data science models that have made success on the Internet-scale level applications like natural language processing, recommendation systems like Netflix, and all the like, do not incorporate scientific theories. Referred as "end of theory" [4, 5], there is a lot of excitement about it among data scientists based on past success. However, these models have drawbacks when it comes to certain scientific domains, including climate science and precipitation. The domain of precipitation in climate is enriched with scientific knowledge, theories that govern rainfall—the major evidences to changing climate by their erratic nature. Hence, it is important to note the significant role that precipitation data can play if analyzed properly through efficient data science models and when wielded can help ascertain the international policy makers to take strong action and eliminate the looming suspicion of changing climate into surety. The authors of this literature review paper discuss the challenges imposed by the nature of climate science data and the traditional data science methods implemented for climate change that is inclusive of precipitation domain [6, 7]. The paper also discusses the concept behind theory-guided data science [8]. The paper surveys the factors that call for a paradigm shift in data analysis [8, 9] and throws light on reasons to pave way for theory-driven data science for precipitation in climate science and hydrology.

2 "End of Theory" in Data Science

In data science, a black-box model approach is a typical analytic approach that completely relies on data. The model as shown in the diagram accounts only for the stimulus and the response from the black-box. The contrary to this model is the white model that even integrates inspection of the theories, logic inside the box.

Black-box model in data science is one of the predictive modeling tools that many data scientists use as black-box model tool-kits. The drawback of black-box data science model can be well understood by the case of Google Flu Trends. Google Flu Trends used data-driven models to estimate the visits to the physicians that were influenza based. It relied on the number of Google Search queries around influenza-based keywords in the USA. The queries' keywords that Google Flu Trends sought were in propensity with the Center for Disease Control (CDC) data [10]. At a cursory view, it is apparently efficacious. The model has, however, received limited success. Eventually due to overestimation of flu propensity based on the number of influenza visits to doctors by more than a factor of two, it failed [7]. As is evident from the black-box technique mentioned earlier, it marks the end of theory. It was quoted by Wired Magazine in 2008 under the End of Science—'The quest for knowledge used begin with grand theories. Now, it begins with massive amount of data. Welcome to the Petabyte age [7, 5].' The key to understanding why Google Flu Trends failed or why black-box approach is insufficient is through knowledge discovery in scientific disciplines. This is also in general helpful in understanding as to why sometimes data science models fail to make success in commercial applications. Evidently, there is a need for a novel model that takes into consideration the deep learning of physics and biochemistry for analysis [7].

2.1 Challenges of Traditional Data Science Models

The black-box model of data science primarily fails because of lack of theory. There is dearth of scientific knowledge with this model, and that results in failure of getting hold of deeper truths from the causal relationship observed. But there is more to it, and the next section shall discuss each factor. The reasons why "black-box" data science model fails are as follows:

(1) Poor Predictive Performance: Ability of the models to predict the outcomes of climate changes becomes poor resulting in unpreparedness.

- Scientific problems are different in nature, with many being complex and under constrains.
- Some problems involve "non-stationary relationships," meaning transient variables. These are the relationships between variables that keep changing with time in sporadic or irregular patterns.
- Often training and testing the data-sets result in inaccurate results as it does not really give the big picture. In other words, that means—it provides a limited representation of the overall phenomena.
- A black-box model relies solely on data, which too is disadvantageous. It leads to spurious relationships that look deceptively good on the training and testing data-sets, but fail when generalizing outside the available data.

(2) Poor Scientific Knowledge: Not incorporating domain knowledge can have negative consequences too, and data and statistics must adhere to what the theories of the domain demand.

- The basic nature of goals differs when it comes to scientific domains. Solely improving predicting performance is not useful or enough unless it also improves our scientific knowledge.
- Non-interpret ability can be viewed as a lack of trustworthiness. For example, a model for medical diagnosis will be more reliable if it can be construed by a medical practitioner.
- Models can serve as building blocks of science only if they are able to communicate to the scientists. Easy translation leads to faster adoption as scientific knowledge.
- End goal is not the output for scientific domains, but the resulting scientific insight is. For example, in biomarkers, detecting genetic traits for a certain disease is not enough—there must be a biological pathway or mechanisms that explain its causal relations (which is how they affect the disease).

3 Climate System—A Data Science Perspective

Climate system includes complex environmental processes in connection with their contribution to global weather. Climate data science studies incorporate data sensed over a short-time span such as a day to very long such as several decades. Monitoring and observing data from these climatic processes is necessary to understand their interactions and how they contribute to grave climate change events like global warming [7]. Precipitation, a sub-domain of climate is a complex combination of climatic processes that deals with various forms of moisture like rain, snow, hail. The goal is to understand the driver behind these processes and then project the impacts (e.g., increasing precipitation and yet the presence of droughts in some regions) statistically. The data from sensing precipitation is huge, and the categories of sources are similar to the sources for any general climate data. Following are the prevailing climate data sources.

3.1 Types of Climate Science Data

Climate science data is divided into three broad classes [6]:

(1) Observational Data—All the observed sensory recordings from in situ ground-based sensors (station-based) to satellites (grid) [6], e.g., NCDC [11], SPARC [12];
(2) Re-analysis Data—In case of dearth of observational sensor data-sets, or the irregular spacing of sensors rendering some interpolation methods function-

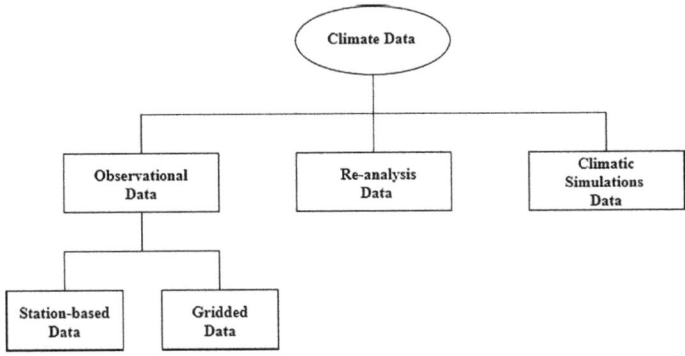

Fig. 1 Types of climate data

less, then use of the comprehensive physical models jointly with the observed sensor recordings, to enhance the process of calculation of missing values or those poor in quality previously—contributes to re-analysis class of climate data sources [6], e.g., MERRA [13], RIO [14];

(3) Climate Simulations—The data from the modeling of dynamically occurring climate phenomenon and processes. Such models simulate physical processes over land, oceans, atmosphere, etc. This greatly contributes to understanding interactions in climatic processes and possibly projecting co-relations among the same [6], e.g., CMIP 5 [12, 15] (Fig. 1).

3.2 Data-Driven Challenges

The traditional data science methods of predictive analysis and causal inference fail in climate science domain. Big data analytics of climate data imposes several challenges to data scientists. A traditional data science method that shall be discusses in the next section and the nature of data that follows explains why data science in climate lags behind others. Besides, climate data and its challenges have not been exposed to a wider community of data science until recently. The following challenges discussed will provide data scientists a perspective for a complete picture of mining of climate data-sets.

Uncertainty and Incompleteness: Due to sensor interference and malfunction of instrument(s) used for recording, the data is likely to be infected with uncertainty and introduces undesired components—also called noise. This condition is acute in regions of heavy atmospheric disturbances, and where the sensors are remotely placed. Atmospheric interference like aerosol, or surface interference like snow and ice if frequently encountered, then data-sets procured from such regions are prone to data uncertainty [6].

Temporal Variability: Ecosystems are bound to have a degree of temporal variability and so will the observed data. For example, the vegetation's data such as greenness normally change over a multi-time scale naturally. But the local and infrequent such as forest fires can induce short-term effects in naturally occurring spatiotemporal effects. It is very important that these temporal variations be handled to avoid detection of spurious patterns [6].

Spatial Auto-correlation: In geography, data is auto-correlated where spatial and temporal propinquity tend to be highly related. Spatial dependence of Earth science data needs to be inculcated in order to obtain physically consistent results. One study in the journal Science [7] identified some correlation (0.9) between the sea surface temperatures of North Atlantic and the Pacific with the forest fires in certain regions of Amazon.

Spatial Heterogeneity: Climate processes naturally exhibit a high degree of variance in space due to changes in geography, topology, and the changes in the climate conditions in different regions on Earth. This calls for the need for local and regional models working on groups of homogeneous locations instead of learning a single global model applicable across all regions of the world [6].

Short Observational Record: Availability of climate data is also a challenge. Some data-sets span a decade or less. But they may have a high spatial resolution. For example, forest fires span over large areas, but on the temporal scale, they are limited. Thus, asking spatiotemporal questions to such data are also limited. In the example mentioned above on Amazon forest fires, the fires data spanned only 10 years on the timescale. Thus, the high correlations (like 0.9 in the example) can occur by random chance and must be scrutinized further [7].

Dynamic Data Representation: Traditional data science methods rely on feeding attribute-value data to machine learning models. But in the case of climate science, there is a raft of processes that cannot be represented in the attribute-value form. A hurricane, for example, does not just appear and then disappear. It rather evolves over a significant spatiotemporal span. And it therefore cannot be represented with binary values. Secondly, data assimilation must consider the heterogeneity of data. One data representation model used in climate science is "climate network" where the nodes represent the geographic locations in the grid, and the edges weight is characteristic of the relationship between the nodes. The study of networks in many applications is assumed to be static networks. However, for climate, intuitive framework is required in order to represent a spatiotemporal network [7].

3.3 Conventional Method Challenges

Traditional data science performs the data exploration through mainly regression or classification. However, climate science requires more of understanding than predicting. This is why black-box models fail. There is interest in predicting individual events, but in climate science, major focus adheres to the system level where emphasis is greater on the interpretability of the model than its flexibility.

The spatiotemporal nature of the data leads to a cross- or auto-correlation between input variables. Thus, analysis methods that make implicit or explicit independent assumptions on the input data variables are rendered futile. Sampling bias is another methodological challenge. Sampling bias is introduced when the model is being trained on biased data or a non-random data-set [7]. This results in poor generalizing of the hypotheses at a later stage. For example, when trying to test the effects on a population, there should be no bias based on age, gender, or even habits. The sample set must be random and approximately consummate representation of the actual population. Another challenge is imposed by defining the problem accurately. Traditional data science has the boon of providing a clear definition of the learning tasks (regression, classification, etc.). In climate science, however, it gets difficult to prepare an objective function. Drought, for example, is quite ambiguous under data perspective. Firstly, there are numerous types of droughts: agricultural, meteorological, and hydrological. There is also a general definition for it, which is about prolonged absence of precipitation. Secondly, even if the definition was decided, how to represent it is unclear, that is, relative or absolute (quantified) terms. If the different methods are employed, then the conclusion would be different. Moreover, unlike traditional data science methods evaluating based on objective metrics, climate science demand more of understanding than mere statistical accuracy. Fundamental difference exists between objective functions in climate science and broader data science. Lastly, there is a need to shift from predictions to causal inference, a novel model that will also be interpreted scientifically.

4 Theory-Guided Data Science

A common problem in scientific domains is representing the relationships between physical variables. Traditionally, scientists resort to theory-based models for representation which are rich in domain/scientific knowledge. Another approach is using data science models. These models make use of learning models, which with the help of input and output variables to determine the relationship between the variables [8]. Theory-based models depend on theory and are used for representing processes that require conceptual understanding using known scientific principles, for example, in the design of satellite trajectories. Data science models on the other hand depend on data, cardinally rely on the information contained in the data, and find applications where amply representative data samples are available such as on internet-scale problems—text mining, object recognition, etc. [8] (Fig. 2).

It can be thus construed that both models individually are limited, irrespective of their other perks. Both are incomplete in the key components: data and scientific theory. This failure is a harbinger for the need of another kind of model called the theory-guided data science model. It is built on the foundations of data science but takes full advantage of the domain theories [16]. It is a next leap in knowledge discovery or exploration to fully realize the vision of a new paradigm. And this new paradigm is the theory-guided data science. It is a novel approach that attempts

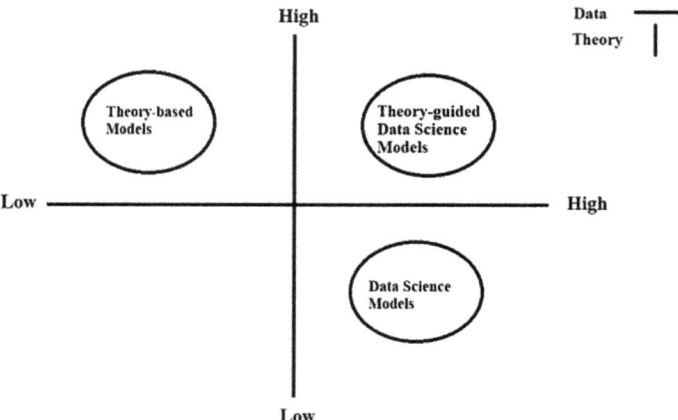

Fig. 2 Contrast drawn between theory-based and data science models

to seamlessly blend the data science and theory-based models and overcome their shortcomings. It is an incorporation or integration of scientific knowledge in data science models. TGDS is an attempt to learn models that are consistent with scientific principles, also known as physically consistent models. This effort surely helps in better generalization than the models poorly based on data. Ideally, it is required to learn a model that shows best generalization performance over any unseen or unfamiliar instance. Practically, models may perform on the training set, but the vision is unaccomplished on a general scale. A number of learning frameworks are found to favor the selection of simpler models that may although have lower accuracy on training sets, they show a better generalization performance. This approach to better generalization is built on the statistical principle of bias-variance trade-off [17]. This trade-off principle forms the heart of many machine learning principles. In engineering, TGDS has found many applications: climate science and hydrology, computational chemistry, biomarker discovery, turbulence modeling.

The overarching vision if theory-guided data science is to include physical consistency as an important component of model performance, other major factors related to performance are accuracy and model complexity. The vision of ensuring scientific interpretability and generalized models will thus be possible. It can be summarized by the following:

Performance Accuracy + Simplicity + Consistency [1]

4.1 Creating Composite Hybrid Models

Composite hybrid models in order to integrate theory and data-sets can be created by two approaches:

(1) **Theory-Based Feature Selection**: In this approach, the domain expertise is commonly used for extracting features. There is a need for automation of domain theory encoding. In this model, principle components not only capture the variance but can also be explained using domain theories. Such a learning model that incorporates domain characteristics finds its applications in Turbulence modeling where the characteristic is scaling, translation, and rotational in-variance. Also, symmetry and near-symmetry are the domain theory for Computational chemistry in data science [16]. For example, the features for precipitation have to be selected considering the region-based characteristics like rainfall received in the area, patterns in rainfall which will involve domain expertise in rainfall. But in snowy regions, it will include both rain and snow patterns as features.

(2) **Feeding Data Science Outputs into Theory-Based Models**: Data science can help fill the gaps where current scientific or domain theory is weak. It can help provide domain theory with intermediate values or predict quantities that improve efficiency of the overall model. This model finds its applications in several domains, for example, Turbulence Modeling. Reynolds-averaged Navier Stokes (RANS) model is the standard tools in modeling the turbulent flow. However, they provide poor approximation in modeling of the complex flows such as swirling, or curvature. The researchers in data science have explored to overcome this problem by means of random forests to predict the discrepancy of RANS model [18]. The random forest model estimates the discrepancy associated in the stress calculation (the cause of the RANS model's approximation to be poor) and further helps augment the RANS model approximation by allowing to scientists to remove the discrepancy so estimated.

4.2 Ensuring Compliance Between Outputs

Domain knowledge or theory can be used to refine the outputs of data science models such that the outputs comply with our understanding of the physical phenomenon. There are several ways of achieving the refinement in the data science outputs such that TGDS leverages even the final stage of the data science models ensuring their consistency with theory. The precipitation science must be obeyed by TGDS outputs. The domain knowledge so used for refinement can be available explicitly or implicitly. Explicitly available theory that refines data science models subsumes model simulations or closed equations. Implicitly available domain knowledge could subsume the latent constraints. There are several approaches to ensuring compliance between outputs and are as shown in Fig. 3.

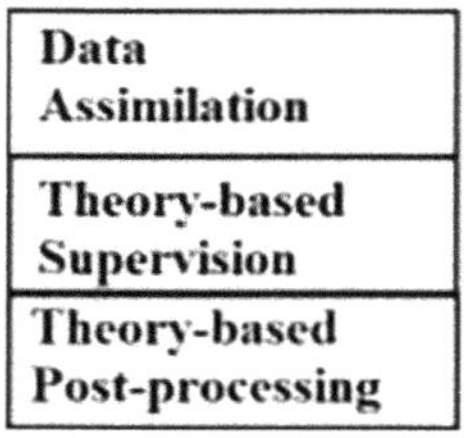

Fig. 3 Approaches to
ensuring compliance
between outputs

| Data Assimilation |
| Theory-based Supervision |
| Theory-based Post-processing |

5 Conclusions

The different challenges of climate data like precipitation, and theory-driven approaches that can be applied for the domain of precipitation have been thrown light on so that it is made aware on a larger scale to the data science community. The "black-box" model is an end of theory in data science. It is insufficient for knowledge discovery in scientific domains of precipitation. Theory-based models rely only on scientific principles and data science models heavily rely on data. There is need to seamlessly blend these two models. The challenges in climate science demand a novel approach that considers the spatiotemporal aspects of climate data and takes into account the very physical nature of the precipitation phenomenon. Theory-guided data science appears as a promising approach to help tackle the big data challenges in climate computing, and hence precipitation where statistical models are not enough. When these models are integrated with domain theory or physics, it can truly help understand the interactions in climate processes involved in precipitation and causing rainfall, snow and also reason out for droughts, etc. Shifting from accuracy metrics to more "physically meaningful" predictive modeling is the key aim of TGDS for climate science. It is a new paradigm in data science that promises knowledge discovery and insights that have significant impact technically as well socially.

As future work, the authors are working on incorporating theory to drive the data science for precipitation considering the factors mentioned in this survey. Scientific knowledge about precipitation forms and various theories such as the ice crystal theory in the domain of precipitation is capable of providing effective insights for understanding the causes, correlations, and consequences of precipitation globally. These prove useful means for scientific knowledge as required by theory-based data science models. Though TGDS for climate science is in incipient state, its themes are at the face of research and innovation that can leap on from the taxonomies discussed in this report.

References

1. Brad B, Jacques B, Michael C, Richard D, Angela H, James M, Charles R (2011) Big data: the next frontier for innovation, competition, and productivity. The McKinsey Global Institute
2. Economist (2010) The data deluge. Special Supplement
3. Szalay A, Bell G, Hey T (2009) Beyond the data deluge. Science 323(5919):1297–1298
4. Halevy A, Pereira F, Norvig P (2009) The unreasonable effectiveness of data. IEEE Intell Syst 24(2):8–12
5. Anderson C (2008) The end of theory: the data deluge makes the scientific method obsolete. Wired Mag
6. A guide to earth science data: summary and research challenges. IEEE https://doi.org/10.1109/mcse.2015.130, 11 Nov 2015
7. Faghmous JH, Kumar V (2014) A big data guide to understanding climate change: the case for theory-guided data science. Big Data 2(3). https://doi.org/10.1089/big.2014.0026
8. Karpatne A, Banerjee A, Ganguly A, Atluri G, Faghmous J, Steinbach M, Samatova N, Shekhar S, Kumar V (2017) Theory-guided data science: a new paradigm for scientific discovery. IEEE Trans Knowl Data Eng 29(10):2318–2331. https://doi.org/10.1109/tkde.2017.2720168
9. Banerjee A, Shekhar S, Faghmous JH (2010) Theory-guided data science for climate change. Published by IEEE Computer Society in November 2014
10. Lazer D, Kennedy R, King G, Vespignani A (2014) The parable of Google flu: traps in big data analysis. Science 343(6176):1203–1205
11. National Climatic Data Center (NCDC). http://www.ncdc.noaa.gov/oa/climate/ghcn-daily/. Last Accessed on 25/2/2017
12. SPARC Data Center. http://www.sparc-climate.org/datacenter. Last Accessed on 25/2/2017
13. Modern-Era Retrospective analysis for Research and Applications (MERRA). https://gmao.gsfc.nasa.gov/merra. Last Accessed on 25/2/2017
14. Reanalysis intercomparison and observations. http://reanalyses.org. Last Accessed on 25/2/2017
15. Meehl GA, Taylor KE, Stouffer RJ (2012) An overview of cmip5 and the experiment design. Bull Am Meteor Soc 93(4):485–498
16. Kumar V (2016) AAAI Symposium, 17 Nov 2016, University of Minnesota
17. Friedman J, Hastie T, Tibshirani R (2001) The elements of statistical learning. Springer series in statistics, vol 1. Springer, Berlin
18. Xiao H, Wu JL, Wang JX (2016) Physics-informed machine learning for predictive turbulence modeling: using data to improve RANS modeled reynolds stresses. arXiv preprint. arXiv:1606.07987

Fusion of Signal and Differential Signal Domain Features for Epilepsy Identification in Electroencephalogram Signals

O. K. Fasil, R. Rajesh and T. M. Thasleema

Abstract Epilepsy is a common neurological disorder and the number of epilepsy patients around the world is increasing in an alarming rate. Identifying and controlling epilepsy is a challenging task. Traditionally, electroencephalogram (EEG) is the most dependable method for the rigorous understanding of epilepsy states. In this paper, a fusion of signal and differential domain features are presented for the effective analysis and identification of epileptic EEG signals. The results of the proposed method for the identification of epilepsy in EEG signal are promising.

Keywords Epilepsy identification · EEG · Fusion of features · Entropy
Signal energy

1 Introduction

In 2015, a video went viral on social media, newspaper, TV channels, and YouTube showing a "drunken" policeman stumbling and eventually crashing down on the floor in the Delhi (India) metro. As a result, several debates are triggered on safety of metro commuters. A day after the video, the policeman was identified and suspended from the duty [1]. Later, it was found that the policeman was not drunk but suffered from a stroke. His dizzy state was the after effect of that stroke [2] and few minutes of that medical condition spoiled his all social life. The same can happen in case of epilepsy patients. The dizziness and fainting that occurs during epilepsy attack will create misunderstanding to normal people, and it will make the patients social life a miserable one.

O. K. Fasil (✉) · R. Rajesh · T. M. Thasleema
Department of Computer Science, Central University of Kerala,
Periye, Kasaragod, Kerala, India
e-mail: fasilok92@gmail.com

R. Rajesh
e-mail: kollamrajeshr@ieee.org

T. M. Thasleema
e-mail: thasnitm1@hotmail.com

© Springer Nature Singapore Pte Ltd. 2018
M. L. Kolhe et al. (eds.), *Advances in Data and Information Sciences*, Lecture Notes
in Networks and Systems 38, https://doi.org/10.1007/978-981-10-8360-0_12

Epilepsy is the most common neurological disorder after Alzheimers, stroke and migraine that has now stamped down many peaceful lives all over the world. Anti-epileptic drugs are not a solution in many cases and only surgical removal of the epileptogenic area would lead to cure [3, 4]. In such cases, the identification of epileptogenic area by manual analysis of recorded EEG signal is the common way in clinical field.

The brain is having a highly complicated structure with million numbers of neurons and its interconnections. The electric potentials (brain signals) that are produced by these neurons are also complicated. Traditionally, these electric potentials/brain signals are captured by placing electrodes over the scalp. There are wide chances for noise in captured signals due to various reasons such as eye blinking, muscle movements. Also, analyzing plenty of data produced by modern EEG machines manually is a time-consuming task, and it requires effective machine learning algorithms. Implanting such algorithms in EEG scanning machines will lead to fast and accurate diagnosis of epilepsy. Also, it will play a vital role in assisting medical doctors. The accuracy/performance of the embedded algorithm is very important, even a small mistake may create worst situations in patient's life. By considering the above facts, we have proposed new method for the identification of epileptic EEG signal.

In this proposed method, the identification of epileptic EEG signal is carried out by the fusion of signal and differential signal domain features and achieved a satisfactory result.

The remaining sections of this paper are organized as follows: Sect. 2 describes some of the previous works related to epilepsy identification. Description of the dataset is provided in Sect. 3. Section 4 deals with the feature extraction. In Sect. 5, the proposed method is explained. Experimental results are discussed in Sects. 6 and 7 concludes the paper.

2 Related Works

Epilepsy identification becomes a challenging and hot topic for research community because of the complexity of the EEG signals. Several methods have been developed in recent years to analyze and identify the epileptic EEG signals. The epileptic region identification is a well-known problem in the literature. Epileptic region identification can be done by classification of focal and non-focal EEG signals. A focal signal is signal which is originating from brain region where epilepsy has affected and non-focal signal is a signal from all non-epileptic regions [4–6].

Out of several methods, a delay permutation entropy (DPE)-based study has been done by Zhu et al. [7], in which authors treated the DPE method as a feature of delay lags of an epileptic EEG signal. DPE is an improved version of permutation entropy which was widely used in EEG classification [8–10]. Authors used SVM classifier with RBF kernel for testing the ability of DPE method and achieved 84% accuracy.

In another work by Sharma et al. [11], authors extracted features in empirical mode decomposition (EMD) domain, which is a well-studied decomposition method in the literature [12–14]. In which, the signal will decompose into intrinsic mode functions (IMF). In their method, they have first decomposed the signal into ten intrinsic mode functions (IMFs) and average sample entropy and average variance of instantaneous frequencies are used as features. LS-SVM classifier classified the signal with an accuracy of 85% while RBF kernel is used. Although the method giving better accuracy, the computational cost which is needed to decompose the EEG signal in a large dataset is extremely higher. In an another work, Sharma et al. [15] studied the capacity of various entropy measures in EMD domain for the classification of focal and non-focal EEG signals and proved that the capacity of the entropy measures for analyzing nonlinear natured signals is promising. This ability of entropy measures is also explored in number of other studies in the literature but not limited to [16–19].

Other than EMD decomposition, discrete wavelet transform-based decomposition was done by Sharma et al. [20] for the classification of focal and non-focal EEG signals. They decomposed the EEG signal upto six levels using Daubechies wavelet and various entropy features are extracted from these DWT coefficients. Authors fed these extracted features to LS-SVM classifier and achieved a classification accuracy of 84%. There are various other studies which depended on the DWT decomposition method for the analysis of epilepsy [21–24].

3 Dataset

A publically available EEG epilepsy dataset collected from Bern-Barcelona EEG database [3, 25] is used in this paper. Dataset contains EEG recordings of five epilepsy patients who are candidates of epilepsy surgery. EEG signals in the dataset are in two classes, namely focal EEG and non-focal EEG, and both class contains 3750 pairs of signal denoted as x and y. Each signal contains 10240 samples and having a length of 20 s. A detailed description of the dataset be read from the paper by Andrzejak et al.

Fig. 1 Pair of focal signals in dataset: **a** Focal x signal, **b** Focal y signal

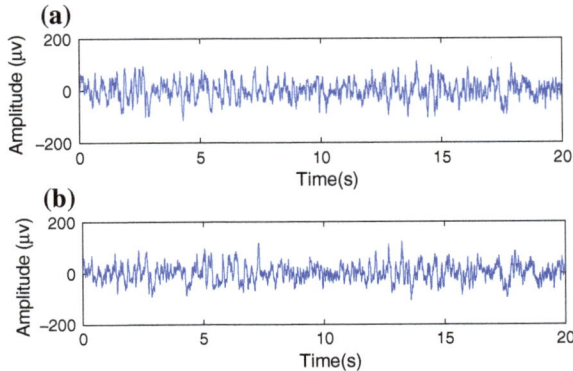

Fig. 2 Pair of non-focal
signals in dataset:
a Non-focal x signal,
b Non-focal y signal

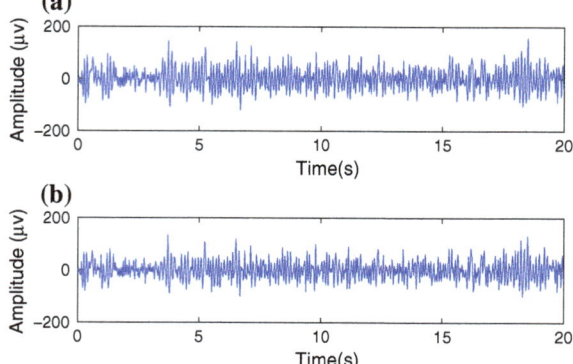

[3]. As used by Zhu et al. [7] and Sharma et al. [20] 50 focal signals and 50 non-focal signals (which is named as Data_F_50 and Data_N_50 in database) are considered for our study. Sample focal and non-focal signals (both x and y) are shown in Figs. 1 and 2, respectively.

4 Features

The two well-known features namely log-energy entropy and signal energy, are used in this study. Experimental results shows that both of these features together can discriminate EEG epileptic signals.

Entropy features are widely used in EEG analysis in many of the studies in the literature [19, 26–28]. As they are highly capable to measure the stochasticity of nonlinear EEG signal and well known for the ability of measuring the randomness in chaotic systems [4]. In this work, we have used log-energy entropy as a feature. Let X be a discrete signal with probability distribution function $p(X)$, then the average log-energy entropy of the segmented signal can be expressed as:

$$X_{LogEn} = \frac{1}{S} \sum_{j=1}^{S} \sum_{i=1}^{N} log_2(p_i(Xj))^2 \tag{1}$$

where i indicates one of the discrete state, j indicates one of the segment, and S is the number of segments.

Signal energy is another feature that we extracted from the signal. Energy is a measure of the strength of the signal. Average energy of a discrete signal $X(n)$ can be expressed as:

$$E = \frac{1}{S} \sum_{j=1}^{S} \sum_{i=-\infty}^{\infty} |X_j(n)|^2 \tag{2}$$

where j indicates one of the segments and S is the total count of segments.

5 Proposed Method

The signals are preprocessed using a Butterworth filter to remove frequencies beyond 60 Hz as they are not influencing classification of epilepsy. In addition to the pair of signal x and y, derived signals like y, x-y, $|x - y|$, $x \times y$ and differential domain signals, namely x', x'', y', and y'' are also considered. If $X = [x_1, x_2, x_3, ...x_N]$ is a single channel discrete signal, then the first-order differential and second-order operations at tth time unit can be defined as:

$$x'_t = \frac{x_{t+\Delta t} - x_t}{\Delta t} \tag{3}$$

$$x''_t = \frac{x'_{t+\Delta t} - x'_t}{\Delta t} \tag{4}$$

where Δt is the difference between adjacent points. In this work, Δt is set as 1. The signals are segmented into ten non-overlapping segments of 2 s duration. Well-known features, namely log-energy entropy and signal energy, are extracted from each of these segments. Average log-energy entropy and average signal energy of all segments of x are considered as the feature for x. Similarly, average log-energy entropy and average signal energy are found from all derived signals in both signal domain (y, x-y, $|x - y|$, $x \times y$) and differential domain (x', x', y' and y'). Features from signal domain and differential domain are then fused together to form feature vector. Thus for each record, 18 features are obtained [i.e., 9 signals (y, x-y, $|x - y|$, $x \times y$, x', x', y', y') $\times$ 2 features (log-energy entropy and signal energy) = 18]. The features are not normalized in this study. These features are passed to SVM classifier for effectual classification. Figure 3 shows the block diagram of the proposed method.

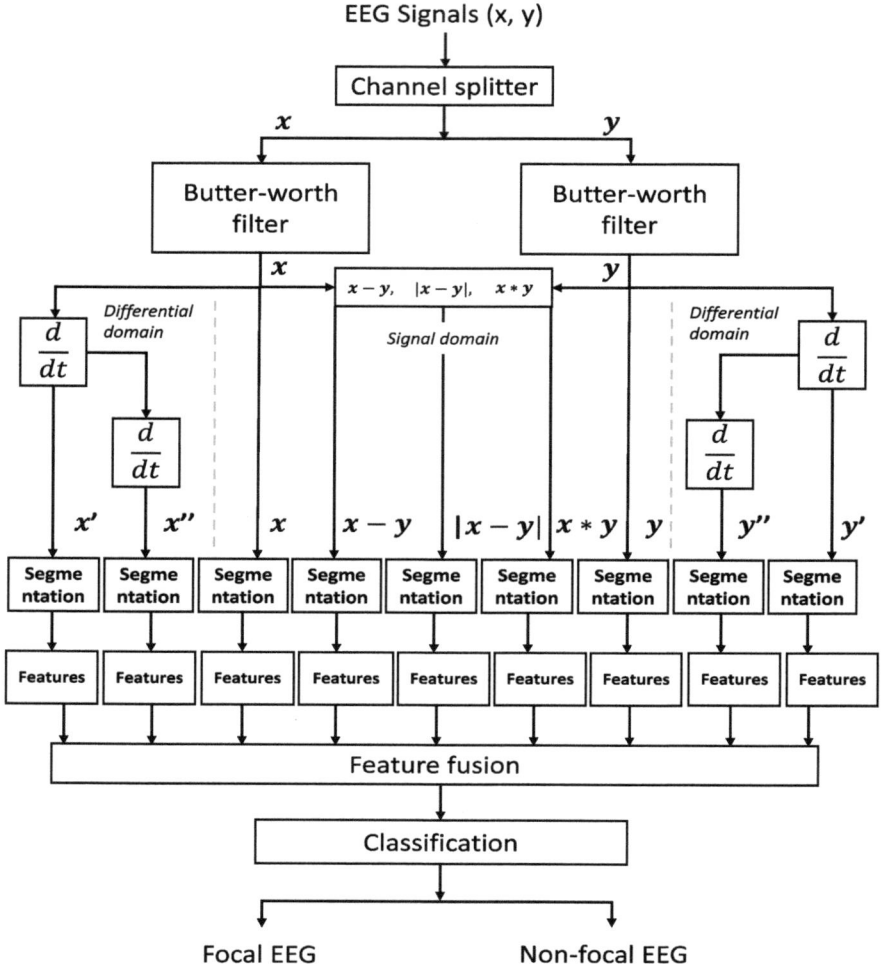

Fig. 3 Overall block diagram of the proposed method

6 Experimental Results and Discussions

As used by Zhu et al. [7] and Sharma et al. [20] 50 focal signals and 50 non-focal signals are used to test the accuracy of our proposed method. Features are extracted from signal domain (i.e., y, x-y, *absolute(x-y)* and $x*y$) and in differential domain (i.e., from x′, x″, y′ and y″). These extracted features are fused together. K-fold cross validation with $k = 5$ using SVM classifier is performed.

The proposed method achieved an average accuracy of 86% (minimum accuracy: 85% and maximum accuracy: 90%). A comparison of the results with existing works is shown in Table 1. Zhu et al. [7] achieved an accuracy of 84% with signal domain features. and Sharma et al. [11, 20] used DWT and EMD domain features and

Table 1 Comparison of classification performance

Method	Domain	Features	Accuracy (%)
Zhu et al. (2013) [7]	Signal	DPE	84
Sharma et al. (2015) [20]	DWT	Entropy features	84
Sharma et al. (2014) [11]	EMD	ASE, AVIF features	85
Proposed method	Signal and differential	Entropy and energy	86

achieved 84 and 85% accuracy, respectively. In proposed work, fusion of signal and differential domain features is used and the results show clear improvements in the accuracy.

7 Conclusion

In this article, features extracted from the signal domain and differential signal domain are fused together for the identification of epileptic EEG signals. Well-known features, namely log-energy entropy and signal energy, are taken as features. The proposed method achieved an accuracy of 86% with SVM classifier, which is better than the results by Zhu et al. [7] and harma [11, 20]. The proposed differential features from signal domain can be extended to other transformed signal domains.

Acknowledgements The authors would like to thank Central University of Kerala for providing research and financial support. The authors would also like to thank the reviewers for their valuable comments to improve the quality of paper.

References

1. IANS: News report on: drunk cop in Delhi metro identified and suspended. Indian Express, http://indianexpress.com/article/cities/delhi/drunk-cop-in-delhi-metro-identified-suspended-bassi/, New Delhi, August 24, 2015
2. Nair HV News report on: That delhi cop in funny viral video was not drunk, he actually suffered a stroke. India Today, http://indiatoday.intoday.in/story/disabled-delhi-cop-who-featured-in-viral-video-knocks-sc-door-seeking-compensation/1/624724.html, New Delhi, March 21, 2016
3. Andrzejak RG, Schindler K, Rummel C (2012) Nonrandomness, nonlinear dependence, and nonstationarity of electroencephalographic recordings from epilepsy patients. Phys Rev E 86(4):046206
4. Das AB, Bhuiyan MIH (2016) Discrimination and classification of focal and non-focal EEG signals using entropy-based features in the EMD-DWT domain. Biomed Signal Process Control 29:11–21
5. Jatoi MA, Kamel N, Malik AS, Faye I, Begum T (2014) A survey of methods used for source localization using eeg signals. Biomed Signal Process Control 11:42–52

6. Liu Q, Chen YF, Fan SZ, Abbod MF, Shieh JS (2016) A comparison of five different algorithms for EEG signal analysis in artifacts rejection for monitoring depth of anesthesia. Biomed Signal Process Control 25:24–34
7. Zhu G, Li Y, Wen PP, Wang S, Xi M (2013) Epileptogenic focus detection in intracranial EEG based on delay permutation entropy. In: AIP conference proceedings (AIP). vol 1559, pp 31–36
8. Labate D, Palamara I, Mammone N, Morabito G, La Foresta F, Morabito FC (2013) Svm classification of epileptic EEG recordings through multiscale permutation entropy. In: The 2013 international joint conference on neural networks (IJCNN). IEEE, pp 1–5
9. Lu L, Zhang D (2016) Based on multiscale permutation entropy analysis dynamic characteristics of EEG recordings. In: 2016 35th Chinese control conference (CCC). IEEE, pp 9337–9341
10. Zhu G, Li Y, Wen PP, Wang S (2015) Classifying epileptic EEG signals with delay permutation entropy and multi-scale k-means. Signal and image analysis for biomedical and life sciences. Springer, Berlin, pp 143–157
11. Sharma R, Pachori RB, Gautam S (2014) Empirical mode decomposition based classification of focal and non-focal seizure EEG signals. In: 2014 International conference on medical biometrics. IEEE, pp 135–140
12. Bajaj V, Pachori RB (2012) Classification of seizure and nonseizure EEG signals using empirical mode decomposition. IEEE Trans Inf Technol Biomed 16(6):1135–1142
13. Meng Q, Chen S, Zhou W, Yang X (2013) Seizure detection in clinical EEG based on entropies and EMD. In: International symposium on neural networks. Springer, pp 323–330
14. Pachori RB, Sharma R, Patidar S (2015) Classification of normal and epileptic seizure EEG signals based on empirical mode decomposition. In: Complex system modelling and control through intelligent soft computations. Springer, pp 367–388
15. Sharma R, Pachori RB, Acharya UR (2015) Application of entropy measures on intrinsic mode functions for the automated identification of focal electroencephalogram signals. Entropy 17(2):669–691
16. Akareddy SM, Kulkarni P (2013) EEG signal classification for epilepsy seizure detection using improved approximate entropy. Int J Public Health Sci (IJPHS) 2(1):23–32
17. Kumar Y, Dewal M, Anand R (2014) Epileptic seizures detection in EEG using DWT-based apen and artificial neural network. Signal Image Video Process 8(7):1323–1334
18. Li P, Karmakar C, Yan C, Palaniswami M, Liu C (2016) Classification of 5-s epileptic EEG recordings using distribution entropy and sample entropy. Front physiol 7:136
19. Li P, Yan C, Karmakar C, Liu C (2015) Distribution entropy analysis of epileptic EEG signals. In: 2015 37th Annual international conference of the IEEE engineering in medicine and biology society (EMBC). IEEE, pp 4170–4173
20. Sharma R, Pachori RB, Acharya UR (2015) An integrated index for the identification of focal electroencephalogram signals using discrete wavelet transform and entropy measures. Entropy 17(8):5218–5240
21. AlSharabi K, Ibrahim S, Djemal R, Alsuwailem A (2016) A DWT-entropy-ann based architecture for epilepsy diagnosis using EEG signals. In: 2016 2nd International conference on advanced technologies for signal and image processing (ATSIP). IEEE, pp 288–291
22. Ibrahim S, Djemal R, Alsuwailem A (2017) Electroencephalography (EEG) signal processing for epilepsy and autism spectrum disorder diagnosis. Biocybern Biomed Eng 38(1):16–26
23. Li M, Chen W, Zhang T (2017) Classification of epilepsy EEG signals using DWT-based envelope analysis and neural network ensemble. Biomed Signal Process Control 31:357–365
24. Sharmila A, Geethanjali P (2016) DWT based detection of epileptic seizure from EEG signals using naive bayes and k-nn classifiers. IEEE Access 4:7716–7727
25. The bern-barcelona EEG database (2012). http://ntsa.upf.edu/downloads/andrzejak-rg-schindler-k-rummel-c-2012-nonrandomness-nonlinear-dependence-and. Accessed 12 July 2017

26. Fergus P, Hignett D, Hussain A, Al-Jumeily D, Abdel-Aziz K (2015) Automatic epileptic seizure detection using scalp EEG and advanced artificial intelligence techniques. BioMed Res Int 2015
27. Lu WY, Chen JY, Chang CF, Weng WC, Lee WT, Shieh JS (2015) Multiscale entropy of electroencephalogram as a potential predictor for the prognosis of neonatal seizures. PloS one 10(12):e0144732
28. Mirzaei A, Ayatollahi A, Gifani P, Salehi L (2010) EEG analysis based on wavelet-spectral entropy for epileptic seizures detection. In: 2010 3rd International conference on biomedical engineering and informatics (BMEI). vol 2, IEEE, pp 878–882

Part II
Smart Computing Techniques

Replica Control Following 1SR in DRTDBS Through Best Case of Transaction Execution

Pratik Shrivastava and Udai Shanker

Abstract Replication is the technique to provide the opportunity of deadline fulfillment of real-time transaction in distributed real-time database system (DRTDBS). Just-in-time real-time replication (JITRTR), on-demand real-time replication (ORDER), on-demand real-time replication with replica sharing (ORDER-RS), periodic transactions using partial replication (PTUDPR), periodicity prediction algorithm, and so on are the replication protocols designed for DRTDBS. None of the algorithms follow one-copy serializability (1SR) correctness criteria for real-time and non-real-time data items. Our main objective is to fulfill strict consistency together with transaction deadline fulfillment. In this paper, the proposed model works on multi-master together with slave sites and uses middleware for following 1SR correctness criteria. In the proposed model, master sites are responsible for executing update and write transactions and slave sides are responsible for executing only read transaction. The proposed model performs better in terms of deadline fulfillment and strict consistency with respect to other replication protocol.

Keywords Replication · Serializability · Strict consistency · Deadline
Multi-Master

1 Introduction

In today's world, distributed real-time database systems (DRTDBSs) have become the need for large number of applications. Time constraint property of DRTDBS makes it different from distributed database. DRTDBS categorizes three types of transactions based on value: hard real-time, soft real-time, and firm real-time. Missing the deadline of hard real-time transaction arises fatal effect in the system. Missing

P. Shrivastava (✉) · U. Shanker
MMM University of Technology, Gorakhpur, U.P., India
e-mail: pratik.shrivastav10@gmail.com

U. Shanker
e-mail: udaigkp@gmail.com

© Springer Nature Singapore Pte Ltd. 2018
M. L. Kolhe et al. (eds.), *Advances in Data and Information Sciences*, Lecture Notes in Networks and Systems 38, https://doi.org/10.1007/978-981-10-8360-0_13

the deadline of soft real-time transaction provides the opportunity to use its value for some time, and missing the deadline of firm real-time transaction gives no value [1]. Consistency constraint, timeliness, and predictability are the major objectives of DRTDBS but there is a trade-off between consistency, timeliness, and predictability. Consistency is compromised for predictability because it enables timeliness. The major source of unpredictability is present in distributed transaction processing. To overcome this problem, replication is used for distributed real-time database [2]. Through full replication, all the data objects are present in all sites such that global transaction can execute locally without coordinating other sites.

DRDTBS uses two types of data objects for transaction execution: real-time data objects and non-real-time data objects. Real-time data object is one whose states get invalid after certain interval of time. Update transaction is periodically executed to maintain its state valid [3]. Non-real-time data object does not have this limitation. It is being modified by application transaction [1, 4]. Application transaction can read/write non-real-time data object but cannot update the real-time data object. Author in paper [3] proposed co-scheduling algorithm for update and application transaction such that consistency of real-time data object together with timeliness of transaction can be fulfilled.

Replication is the technique for increasing the availability of services and executing real-time transaction within its deadline. There exist different variants to handle replica such as primary copy, majority protocols, ROWA, ROWA-A, and quorum consensus. Replication protocol is used to maintain the consistency of data objects in the replicas. It can be categorized as eager or lazy and primary copy or update anywhere. Eager propagation assumes that all replicas should get updated before getting commit and lazy propagation assumes that transaction gets commit first and propagation of message is done later. In primary copy-based replication, there is only one master and others are slave. In update anywhere, there are more than one number of masters. Different authors have proposed different replica control algorithm for maintaining the consistency [5]. But to the best of our knowledge, none of the authors follows one-copy serializability criteria for real-time transaction execution in DRTDBS. So, this paper proposes replica control algorithm based on one-copy serializability criteria. The contribution of this paper is as follows.

1. Proposed algorithms work in different sublayer of middleware with 1SR correctness criteria.
2. Middleware performs transaction data analysis, conflict detection and propagation.

2 Literature Review

There has been extensive research done by the researchers in replication for DRT-DBS. In the paper [6], author integrates replication protocol with real-time scheduling. The replica control algorithm was based on majority consensus scheme and follows epsilon correctness criteria. Due to epsilon serializability, some inconsistency must be tolerated for read-type transaction. In the paper [7], author proposes replication algorithm for distributed real-time object-oriented database. JITRTR algorithm creates replication transaction which copies the required data object from remote to local such that transactions get the updated value. But this replication algorithm works in static environment where data object location and user transaction requirement were known in advance. This algorithm does not deal with dynamic request and is not suitable for DRTDBS. In the paper [8], author has proposed a replication algorithms ORDER and ORDER-S which create the replica based on the request by the incoming transaction and its update frequency is decided based on the data freshness requirement by the transaction. This replica updation based on request makes short duration real-time transactions to miss their deadline which may cause transactions to be aborted for firm real-time transaction or cause fatal effect to system for hard real-time transaction. In the paper [9], author uses dynamic replication together with partial replication for dynamic workload in real-time distributed database. Here, author has proposed two replica control algorithms; one is based on non-static periodic transaction arrival termed as periodic transactions using partial replication (PTUDPR), and another is based on random transaction arrival termed as periodicity prediction algorithm. Due to the unpredictability of dynamic request and partial replication, sometimes real-time transaction can miss their deadline. In the paper [2], author has proposed a continuous convergence protocol for distributed real-time database. This protocol has three main terms: local consistency, local predictability, and eventual global consistency. Local consistency and predictability are the need of real-time transaction for execution on the local site, and eventual global consistency allows read transaction to read inconsistent data. If same data objects get concurrently updated by more than two transactions, then instead of being rollback, it is being resolved through forward conflict resolution mechanism. This algorithm has to tolerate inconsistency for read transaction. In the paper [10], author proposes a real-time replication control protocol (RT-RCP) for distributed real-time database. RT-RCP protocol focuses on maintaining the consistency of non-variant (i.e., non-real-time) data objects. The main idea behind RT-RCP is that the owner site executing the transaction will update as many replicas coming under transaction deadline and the remaining replicas will be updated after commitment of transaction. In the paper [11], author has proposed a configuration of replication technique for increasing the availability and performance in real-time database. Replication protocol used to manage non-real-time data cannot be used to manage real-time data. In this paper, the author focuses on maintaining the consistency of non-real-time data. To fulfill the time constraint of real-time transaction, the author binds the DLR-ORECOP algorithm with commit protocol, so that, maximum transaction can be completed within their deadline. In the paper [5], author increases the scalability of DRTDBS through using virtual replication, state transfer propagation in place of operation transfer, and on-demand updating scheme. This paper follows optimistic replication protocol.

After going through all these papers, none of the authors in my best of knowledge focuses on strict correctness criteria (1SR) for DRTDBS.

Pessimistic replication guarantees that transaction during execution does not read the stale data. It creates illusion to user that there exists single available copy of data. Previous pessimistic replication protocols are suitable for LAN because failure and latencies are small. Pessimistic replication is not suitable where network is very slow and node is frequently disconnected. It shows poor performance when the nodes in system increases [12]. Due to all these drawbacks, pessimistic replication protocol shall not to be used for real-time database system. But today, due to the presence of high-speed network and total order broadcast, pessimistic replication protocol becomes capable to be utilized in real-time database system. High-speed network makes the guarantee that the message send by the sender will be received by the receiver in a predictable amount of time, and total order broadcast ensures that the order of message send by the sender will be executed by the receiver in the same order.

There are various correctness criteria which can be followed for replicated databases serializability [13]. One-copy serializability means that the effects of concurrent transactions performed on different replicas have an ordering which is equivalent to an ordering obtained when the same number of transactions is performed sequentially in a single centralized database. Serializability does not allow the alteration order of transaction execution even if they produce the consistent result [14, 15].

In this paper, middleware is proposed which implements replication algorithm that follows 1SR for application and update transaction execution. By following 1SR, none of read transaction has to read inconsistent data. Pessimistic replication protocol allows 1SR but it may cause concurrent transaction requiring same data objects to wait because of locking data objects by another transaction. So, in this paper we will use the approach of paper [16] which improves the response time of transaction through thread-to-metadata policy. The proposed approach will try to complete transaction execution together with all replica updation within transaction deadline. Due to concept of paper [16], concurrent transaction waiting for same item will get decreased.

The organization of the rest of the paper is as follows. Section 2 deals with literature review. Section 3 describes system model. Description of protocol has been done in Sect. 4. Section 5 describes summarized contribution. Section 6 concludes the paper.

3   System Model

This system model is composed of N number of masters which are connected through high-speed network, and masters are fully replicated which provides the opportunity to the master to handle non-static, dynamic, and periodic transaction. Periodic transaction was updating the value of real-time data item such that transactions get the updated value for their execution. Dynamic transactions need unexpected data items for their execution, and non-static transactions were coming non-statically on the master site but their data items requirement were already known. Every master consists of two types of data item: real and non-real. These both data items may be used by dynamic and non-static transaction. Real-time data items are attached with a validity interval which states that its validity gets elapsed if it is not updated before its validity time expires, whereas non-real-time data item does not have such limitation.

The proposed system architecture consists of master site, slave site, clients, and middleware shown in Fig. 1, and detailed description is given below.

A. Master site: The master site is responsible for executing update transaction and write transaction. Update transactions were of very less deadline, whereas write transaction can be of less or large deadline. Master site holds both data item. It is also responsible for forwarding the read transaction to slave node based on load status set by the slave. The execution of transaction in master site follows thread-to-metadata policy in place of thread-to-transaction because it improves the response time of transaction as defined in the paper [16].

B. Slave site: This site is responsible for executing only read transaction. All the data items required by the transaction for execution are available on the site. Slave sites fetch the updated data item value from the master periodically for getting updated. The execution of read transaction on the slave site and update, write transaction on the master site makes the master CPU load shared.

C. Client: Clients are the users who want the services from the system, and they are forwarding the request in the form of SQL query to the system. Request can be of type read or write.

D. Middleware: This middleware is the link between client and server. This middleware will accept the request, perform some functions, and based on the result, it will forward the request to server for completion. This middle layer consists of three sublayers where each sublayer has its own functionality for fulfilling the 1SR correctness criteria. The detailed description of each sublayer is given below.

(a) Transaction Data Analyzer: This layer identifies the data item requirement from the transaction. Basically, transaction is a collection of set of com-

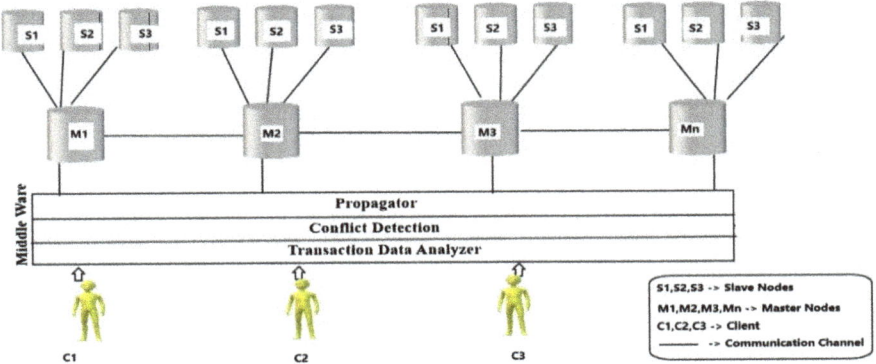

Fig. 1 1SR multi-master architecture for DRTDBS

mands where each command can be categorized into read, write, or update and for executing the command, data items are required. So, from the collection of commands, this layer identifies the set of required data items. This layer also identifies the deadline of the transaction. After identifying deadline and set of data items, this information is passed to upper layer for conflict detection.

(b) Conflict Detection: This layer identifies the conflict between different transactions. The detection of conflict is done on the basis of three conditions.

   (i) Operations should belong to different transaction.

   (ii) Same data item should be accessed.

   (iii) Operations can be combination of read–write, write–read, or write–write.

This layer records the history of completed transaction, executing transaction, and forthcoming transaction. It checks the conflict between executing and forthcoming transaction. If there is conflict between them, then forthcoming transaction may be restarted in future based on commit or abort of executing transaction. This layer checks the serializability of all transactions.

(c) Propagator: This layer will total order broadcast update, write transaction to all master such that they all reach in the same order. Read transaction are unicast to only master site randomly such that master will forward this request to eligible slave having less load.

The main role is played by the middleware in this system. All three sublayers of middleware perform different functionality for following the serializability correctness criteria (Table 1).

4 Proposed Algorithm

This proposed algorithm "Identify List of Data Items from Transaction" works in transaction data analyzer sublayer, and it is executed whenever any transaction enters for execution. In this algorithm, loop is executed until it founds commands in the transaction. Information about extracted data item is stored in queue. The finalized resultant queue will be forwarded to conflict resolution sublayer.

Algorithm followed in transaction data analyzer is described below.

Table 1 List of symbols and definitions

Symbol	Definition
Tre	Requested transaction
C	Command
D	Data item
Qi	Requested queue
Tr	Read transaction
Tu	Update transaction
Tw	Write transaction
Dw	Deadline of write transaction
Du	Deadline of update transaction
Dr	Deadline of read transaction
R	Result
CNT	Counter
RSTO	Restart option
M	Master site
Sl	Slave site
S	Schedule
T	Waiting time

```
Result: List of data items
Requested Transaction Tre;
while C exist in Tre do
    Identify D from C;
    Store D in Qi one by one;
end
Identify type Tr, Tu;
Enqueue transaction type in Qi;
Finalized Qi is send to upper layer;
```

Algorithm 1: Identify List of Data Items from Transaction

This proposed algorithm "Identify Conflicted Transaction" works in conflict detection sublayer. In this algorithm, deadlines of update transaction and read transaction are compared whether they are coinciding, if they then check data conflict. This algorithm will mark that conflicted read transaction with RSTO options which marks the master that this read transaction will be restarted or not that depends upon whether update transaction or write transaction is committed or aborted.

Algorithm followed in conflict resolution is described below.

```
Result: Serialized Transaction
Select Tu or Tw from Qi;
Identify Du, Dw and Dr of Tu,Tw and Tr respectively;
while If Du or Dw coincides with Dr do
        Compare each D of Tu with D of Tr;
        Store compare result in R;
    if R is true then
                Tr conflicted with Tu;
                CNT incremented by 1;
    end
end
if CNT>0 then
    Mark Tr with RSTO;
else
    Don't Mark Tr with RSTO;
end
Forward the Tr to next upper layer;
Finalized queue is send to upper layer;
```

Algorithm 2: Identify Conflicted Transaction

This proposed algorithm "Identify Schedule Following Serializability" works in propagator sublayer. In this sublayer, schedule for concurrent execution of transaction is verified. The generated schedule for concurrent transaction follows the serializability criteria.

Algorithm followed in propagator is described below.

```
Result: Broadcast or Unicast Transaction
foreach T do
    Create S following 1SR with Deadline;
end
if S is Successful then
  foreach T do
    if Check T type then
            Tu or Tw will be total order broadcast;
    else
            Tr will unicast to any M;
    end
end
else
        Regenerate S;
End
```

Algorithm 3: Identify Schedule Following Serializability with Deadline

Fig. 2 Load table

Master Site ID	Load Status
SL1	Overload
SL2	Underload
SL3	Overload
● ● ●	● ● ●
SL(n-1)	Underload
SLn	Overload

Master Site

All master sites are responsible for executing update and write transaction. They all execute the set of update and write transaction in the same order as per finalized by the propagator. Master site is also responsible for selecting the slave site based on the load status. Slave site periodically sends their load status to their respected master. This load status information is stored in the table which will be used by the master site. The load table structure is given in Fig. 2.

Algorithm used by the master for selecting the slave is given below.

Result: Selected M
foreach Sl in Load Table **do**
    if Sl Load Status is UNDERLOAD **then**
            Select Sl for Tr;
            Forward the Tr in Sl;
            From selected Sl;
        if Tr is marked with RSTO **then**
            if deadline of Tr not violated **then**
                Wait for T (T<Tr deadline);
                Fetch and Operate on updated value;
                Respond to client with result;
            else
                Abort the transaction;
        else
            Work on updated value;
            Respond to client with result;
    else
            Check for next Sl;
    end
end

Algorithm 4: Slave Selection

5 Summarized Contribution

The proposed system will be performed better due to the following reasons.

1. In the proposed system, there will be large probability of transaction deadline fulfillment because master sites only have to execute write, update transaction based on its priority value, whereas slave sites are more as compared to master sites and they have to execute only read transaction.
2. In the proposed system, we use thread-to-metadata policy in place of thread-to-transaction policy [16] because response time of transaction improves through following thread-to-metadata policy. This policy will help the transaction to complete within its deadline.
3. In the proposed system, we use high-speed network and total order broadcast for message propagation because in highly speed network, message sent from the master site to slave site will be received in a predictable amount of time. Total order broadcast guarantees that the execution order of message on slave site is same as it is sent by the master. Due to high-speed network and total order broadcast, all slave sites will reach to same state.
4. In the proposed system, read and write transaction execute on different place locations. Due to this, CPU load on the master site and slave site decreases.

5. The scalability of the proposed system does not get compromised even if number of read transaction increases because slave sites are in large number as compared to number of master sites for executing read transaction.

6. The proposed model follows strict correctness criteria because read transaction can work only on consistent data. If read transaction is conflicted with ongoing write or update transaction then in that case if it is possible to wait then read transaction will wait for getting updated value and then execute, otherwise read transaction will abort.

7. The performance of the system will not decrease because of different place location of transaction execution.

8. Proposed system provides the property of high availability because all sites are fully replicated and even if one or more than one site got failed, then also the request will be fulfilled by the nearest available server.

9. Proposed system provides the illusion of single server even if some sites got failed because all sites are fully replicated.

6 Conclusion

In this paper, the proposed system uses multi-master together with slave sites for DRTDBS. The proposed system consists of middleware which follows 1SR correctness criteria for transaction execution. Middleware consists of three sublayers where each sublayer has its defined functionality. In the proposed system, read transaction is executed on the slave sites and write, update transactions are executed on the slave sites such that it provides the opportunity to fulfill the deadline requirement of transactions. The proposed system works on the consistent values and returns consistent result. The proposed algorithm considers both real-time data and non-real-time data and will perform better in terms of deadline and strict consistency fulfillment.

References

1. Ramamritham K, Son SH, Dipippo LC (2004) Real-time databases and data services. Real-time Syst 28(2–3):179–215
2. Gustavsson S, Andler SR (2005) Continuous consistency management in distributed real-time databases with multiple writers of replicated data. In: Proceedings 19th IEEE international parallel and distributed processing symposium, 4 Apr 2005, IEEE, p 8
3. Han S, Lam KY, Wang J, Son SH, Mok AK (2012) Adaptive co-scheduling for periodic application and update transactions in real-time database systems. J Syst Softw 85(8):1729–1743
4. Shanker U, Misra M, Sarje AK (2008) Distributed real time database systems: background and literature review. Distrib Parallel Databases 23(2):127–149
5. Salem R, Saleh SA, Abdul-kader H (2016) Scalable data-oriented replication with flexible consistency in real-time data systems. Data Sci J 17:15
6. Son SH, Zhang F (1995) Real-time replication control for distributed database systems: algorithms and their performance. In: DASFAA 1995, 11 Apr 1995, pp 214–221

7. Peddi P, DiPippo LC (2002) A replication strategy for distributed real-time object-oriented databases. In: Proceedings of fifth IEEE international symposium on object-oriented real-time distributed computing, ISORC 2002, IEEE, pp 129–136
8. Wei Y, Aslinger A, Son SH, Stankovic JA (2004) ORDER: a dynamic replication algorithm for periodic transactions in distributed real-time databases. In: 10th international conference on real-time and embedded computing systems and applications, RTCSA, 25 Aug 2004
9. Aslinger A, Son SH (2005) Efficient replication control in distributed real-time databases. In: The 3rd international conference on computer systems and applications, ACS/IEEE 2005, IEEE, p 34
10. Haj Said A, Sadeg B, Amanton L, Ayeb B (2008) A protocol to control replication in distributed real-time database systems. In: Proceedings of the tenth international conference on enterprise information systems, vol 1, ICEIS 2008, pp 501–504. ISBN 978-989-8111-36-4
11. Said AH, Sadeg B, Ayeb B, Amanton L (2009) The DLR-ORECOP real-time replication control protocol. In: IEEE conference on emerging technologies and factory automation. ETFA 2009. 22 Sep 2009, IEEE, pp. 1–8
12. Saito Y, Shapiro M (2005) Optimistic replication. ACM Comput Surv (CSUR) 37(1):42–81
13. Ouzzani M, Medjahed B, Elmagarmid AK (2009) Correctness criteria beyond serializability. In: Encyclopedia of database systems 2009, Springer, US, pp. 501–506
14. Ruiz-Fuertes MI, Munoz-Escoi FD. A consistency-based specification for the one-copy serializability variants
15. Ruiz-Fuertes MI, Munoz-Escoi FD (2011) Refinement of the One-copy serializable correctness criterion. Tech. Rep. ITI-SIDI-2011/004, Instituto Tecnológico de Informática, Valencia, Spain
16. Mokhtar HM, Adel N (2012) Transaction processing using thread-to-metadata. In: Proceedings of the 16th international database engineering and applications symposium, 8 Aug 2012, ACM, pp 230–234

FPGA Implementation of a Fast Scalar Point Multiplier for an Elliptic Curve Crypto-Processor

Satvik Maurya and Vaishali Ingale

Abstract This paper presents a fast scalar point multiplier for an elliptic curve crypto-processor in the field GF $\left(2^{163}\right)$. Elliptic curve-based cryptographic algorithms have been in wide use since the early 2000s after being introduced in 1986. With the ever-increasing need for information security, it is essential for systems to perform the required operations in a fast and efficient manner. In this work, a hybrid type Karatsuba multiplier has been used for fast field multiplications and a dedicated inverter module based on the extended Euclidean algorithm is used for fast field inversions. The point multiplication is performed using the standard double-and-add algorithm for which the point doubling and point addition are done using standard projective coordinates. The use of the fast multipliers and field inverters makes the implementation a fast one as compared to other high-performance implementations that have been reported over the years, at the cost of increased resource usage. The results obtained, however, justify this increased resource usage as the point multiplication is the most time-intensive operation in the encryption and decryption process of elliptic curve cryptography.

Keywords FPGA · Karatsuba multiplier · Extended Euclidean algorithm · Elliptic curve cryptography (ECC) · Scalar point multiplication

1 Introduction

Cryptography, in its most classical form, has been around for thousands of years. From Julius Caesar, who used a simple substitution cipher to relay messages to his generals, to the Second World War, in which the breaking of the German cipher

S. Maurya (✉)
Delhi Technological University, Delhi, India
e-mail: satvik_bt2k14@dtu.ac.in

V. Ingale
College of Engineering, Pune, India
e-mail: vvi.extc@coep.ac.in

© Springer Nature Singapore Pte Ltd. 2018
M. L. Kolhe et al. (eds.), *Advances in Data and Information Sciences*, Lecture Notes in Networks and Systems 38, https://doi.org/10.1007/978-981-10-8360-0_14

hastened the end of the war, cryptography and subsequently cryptanalysis have been a part of society for a very long time. Till the 1970s, cryptography was an obscure field, used primarily by the military and spy agencies. With the advent of modern computers, and along with it the invention of Data Encryption Standard (DES) and most importantly the RSA public-key cryptography algorithm, the field of cryptography was brought to the public domain. With the increasing need to protect data from unauthorized usage along with the ever-increasing computational power available, new algorithms are constantly being developed. Elliptical curve cryptography is an algorithm which has gained popularity in recent times, even though it has been around since the 1980s.

ECC is a public-key cryptographic algorithm first developed in 1985 by Neal Koblitz and Victor S. Miller, and it has been standardized by [1–3]. Like the RSA, ECC functions in a finite field, which can be a prime field represented by GF(p) or a binary field GF(2^m), where p and m are prime numbers. The chief reason for the popularity of ECC is the smaller key size required to encrypt data as compared to the RSA algorithm. For example, the security offered by a 1024-bit key in RSA can be offered by a 160-bit key in ECC. Both RSA and ECC are used in applications where a secure communication is to take place between two entities, and thus, for applications such as IoT, Near-Field Communication (NFC), Bitcoin, their use is extensive. The smaller key requirement of ECC thus enables it to have lesser hardware, making it suitable for mobile applications.

The functionality of ECC is based on the manipulation of points present on the elliptic curve, the most important being the scalar point multiplication $Q = kP$ which is the addition of P to itself $k - 1$ times. The security of ECC is thus based on this manipulation, that given two points P and Q on the elliptic curve; it is difficult to calculate the value of k. This problem is known as the Elliptic Curve Discrete Logarithm Problem (ECDLP).

The relative ease with which software applications can be compromised has resulted in the general popularity of hardware approaches towards cryptography. Dedicated crypto-processors not only perform the encryption faster, they also provide security at a physical level from potential attackers as they cannot be compromised without direct physical access to them. Hardware approaches towards ECC can be broadly classified into two main categories based on the finite field used. Approaches which use the prime field GF(p) are few as compared to the approaches which use binary fields GF(2^m) [4]. This skew is due to the difficulties involved with large prime numbers on hardware as well as the ease with which binary fields can be represented on hardware. As the scalar point multiplication is the most hardware and time-intensive operation involved in any elliptic curve application such as the elliptic curve Diffie–Hellman (ECDH) key protocol or the Elliptic Curve Digital Signature Algorithm (ECDSA), it is this operation that has been the prime area of focus for crypto-processor systems. A high-speed crypto-processor implementation using the Montgomery Point Multiplication Algorithm was presented in [5, 6] which used pipelining as well as multiple multipliers and squarers for reducing the latency of operations. Dual field approaches which implement both prime and binary fields are also becoming popular, such as the architecture proposed in [7]. In general, most papers on the hardware implementation of ECC alternate between the various point multiplication algorithms [8–11] and the inversion algorithm used, with the Itoh-Tsujii being the most efficient algorithm in this case [4].

This paper presents an implementation of an elliptic curve crypto-processor in the finite field of GF (2^{163}). The point multiplication has been accomplished using standard projective coordinates, and the extended Euclidean algorithm has been used for the inversion. To perform quick field multiplications, a hybrid Karatsuba multiplier has been used. The structure of the paper is as follows: Sect. 2 presents the basics of ECC as well as the architecture of the modules used in this work, Sect. 3 presents the implementation and results, and conclusions are put forth in Sect. 4.

2 Methodology

2.1 Elliptic Curve Cryptography Preliminaries

The elliptic curve E for a binary field GF(2^m) is given below:

$$E\colon y^2 + xy = x^3 + ax + b \tag{1}$$

where $a \in [0, 1]$ and $b \neq 0$. Any operation on a point that satisfies the curve E follows a specific set of group laws which govern how the operation can be performed. These group laws differ according to the field in which the elliptic curve is defined.

Elliptic curves have a wide range of modular arithmetic operations associated with them. These include the field multiplication, field squaring, field reduction, field addition and the field inversion operations. As mentioned earlier, the scalar point multiplication is the most important operation in ECC. The scalar point multiplication involves repeated point doublings and point additions which in turn require the field arithmetic operations mentioned above. Depending on the algorithm used for the point multiplication, field inversion can be required for every point doubling and addition or just once for every point multiplication. This variability comes from the use of coordinate systems such as standard projective coordinates and the Lopez-Dahab projective coordinates that differ from the one in which the curve E is defined in. The field arithmetic as well as the group laws governing elliptic curves are given in detail in [12].

2.2 Proposed Design

The architecture for the proposed arithmetic unit for implementing a point multiplication for ECC in GF (2^m) is shown in Fig. 1. A hybrid Karatsuba multiplier has been used for the field multiplication along with a field inverter based on the extended

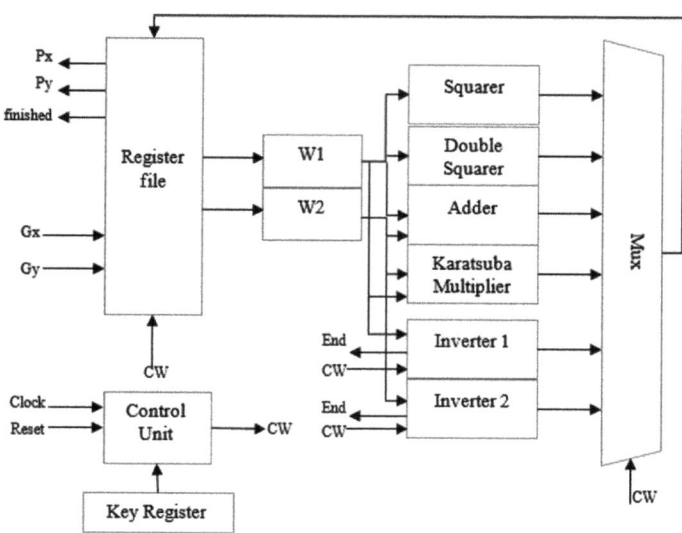

Fig. 1 Architecture for the arithmetic unit

Euclidean algorithm. The dedicated hardware made for field inversions guarantee a fast inversion process at the cost of increased area usage.

The register file contains the registers used for storing the values of the intermediate as well as final coordinates of the point multiplication process. The registers W1 and W2 are the two working registers that are loaded with the values that need to be operated on. The working registers form the input to the arithmetic blocks, namely the squarer, double squarer, adder, multiplier and field inverters. Working register W1 is used to provide the singular inputs to the adder, squarer, double squarer and Inverter1 as well as the dual inputs to the multiplier along with working register W2. W2 also provides the input to the second inverter module, Inverter2. It should be noted that in a field of $GF(2^m)$, all registers and wires are m-bits wide.

The algorithm that has been used for the scalar point multiplication is called the left-to-right double-and-add algorithm. Starting from the MSB of the key, every bit is checked whether it is a 0 or a 1. Depending on the value of the current bit, a point doubling and a point addition are performed on the points. The algorithm is given in Fig. 2.

Thus, if the present bit of the key is 1, a point addition is performed after the point doubling. The total length of the key in this case will decide the worst-case execution time of the entire point multiplication.

The point addition and doubling are part of the group law which governs the elliptic curve being used. When standard affine coordinates (X, Y) are being used, the point doubling and point additions require a field inversion every time they are executed. Because the field inversion is the most time-intensive operation of the point multiplication process, a better alternative is to use a different coordinate system which requires a field inversion only once during the entire point

Left-to-right double and add algorithm for scalar Point Multiplication

Input: Generator point $G \epsilon E(GF(2^m))$, key $k = (k_{l-1}, \ldots \ldots k_0)$.

Output: Product $P = kG, P \epsilon E(GF(2^m))$

1. $P \leftarrow \infty$
2. *For i from l − 1 to 0*
   2.1. $P \leftarrow 2P$
   2.2. *If* $k_i = 1, P \leftarrow P + G$
3. *Return(P)*

Fig. 2 Algorithm for scalar point multiplication

Fig. 3 Point arithmetic using standard projective coordinates

Point Doubling and Point addition using Standard Projective Coordinates

$E: y^2 + xy = x^3 + ax + b, a = 1, b = 1$

Point Doubling: $2(X_1, Y_1, Z_1) = (X_3, Y_3, Z_3)$

1. $Z_3 = X_1^2 . Z_1^2$
2. $X_3 = X_1^4 + b.Z_1^4$
3. $Y_3 = b.Z_1^4.Z_3 + X_3(a.Z_3 + Y_1^2 + b.Z_1^4$

Point Addition: $(X_1, Y_1, Z_1) + (G_x, G_y, 1) = (X_3, Y_3, Z_3)$

1. $A = Y_2.Z_1^2 + Y_1$
2. $B = X_2.Z_1 + X_1$
3. $C = Z_1.B$
4. $D = B^2.(C + a.Z_1^2)$
5. $Z_3 = C^2$
6. $E = A.C$
7. $X_3 = A^2 + D + E$
8. $F = X_3 + X_2 Z_3$
9. $G = X_3 + Y_2 Z_3$
10. $Y_3 = E.F + Z_3.G$

multiplication process. One such system is the standard projective coordinate system which introduces an additional coordinate Z which reduces the point doubling and addition operations to field multiplications, squaring and addition. The projective coordinates can be converted back into affine coordinates by using a suitable transformation. A method of point doubling and point addition using standard projective coordinates is given in [13]. During point addition, the point P is added to the starting point G. Thus, the generator points (G_x, G_y) in affine coordinates can be converted into the projective coordinate $(G_x, G_y, 1)$. This can be verified using the projective to affine conversion $(X, Y, Z) \rightarrow (X/Z, Y/Z^2)$. Thus, the final coordinates in P can be converted into affine coordinates to obtain the points (P_x, P_y). The algorithm for point doubling and point addition is given in Fig. 3.

The field squaring, double squaring, multiplication and addition are performed using entirely combinational circuits which thus enable these operations to be performed in a single clock cycle. The addition, squaring and double squaring can be done using simple XOR operations. The various modules including the control logic are explained below:

I. *The Karatsuba Multiplier*: The Karatsuba multiplier is one of the many alter-
natives available for designing a finite field multiplier. The efficiency of the
Karatsuba multiplier, especially for large number multiplication, has made it
one of the most popular choices for the finite field multiplier module in ECC
cryptosystems. The Karatsuba multiplier used in this paper differs from the traditional recursive Karatsuba multiplier that requires operands with a bit length
of an integral power of 2. To accommodate the fields and word lengths required
in ECC, a hybrid Karatsuba multiplier is used as presented in [14]. The 163-
bit multiplication is split into two multiplications with lengths 81 and 82 bits,
respectively. These are further split into 40- and 41-bit multiplications, and
the remainder of the tree can be grown likewise. The last layer of multipliers, namely the 20-bit and 21-bit multipliers, can be implemented using simple
AND and XOR logic. The properties of polynomial arithmetic in GF(2^m) do
not allow the generation of any carries during any arithmetic operation, and
thus, combining the multipliers together to obtain the final product requires
simple XOR operations. It should be noted that a field multiplication of $2m$-bit
numbers will result in a product with ($2m - 1$) bits. The modular reduction of
this product according to the irreducible polynomial of that field is done using
a reduction module that is present with the multiplier. This reduction unit has
not been shown in the diagram, but it converts the ($2m - 1$) bit product to a
m-bit product. The multiplier tree for the field GF (2^{163}) is shown in Fig. 5.

II. *Finite Field Inversion*: The finite field inversion has been performed using the
extended Euclidean algorithm for binary fields, as shown in Fig. 4. The extended
Euclidean algorithm works on the principle of continuously dividing the polynomial until the remainder is 1. While the area requirement of such an inversion
process is larger than other approaches such as the Itoh-Tsujii [15], it has the
advantage of being faster. The inverter block shown in Fig. 1 also has the control
word input from the control unit and an "End" output which signals the end of
the inversion process. The control unit reads this output.

III. *Control Unit*: The control unit controls which register values are transferred to
the working registers as well as the selection of outputs of the various arithmetic modules. The state diagram for the control unit is shown in Fig. 6. The
control unit issues a control word (CW) which is specific to the present state
of the control unit. The states are broadly classified in four different types: the
initial state, which initializes all registers and is active until the active low asyn-
chronous reset is enabled; the point doubling and point addition states which
are switched according to the present bit of the input key; the projective to affine
conversion state, which is entered when the point multiplication is complete.
The projective to affine state issues the control word for starting the inversion
process in the inverters and reads the "End" output of both inverters. The inverter
outputs are transferred to the register file only after both inverters have completed their operations. Once the point multiplication has been completed, the
control unit asserts the "finished" signal which indicates the end of the point
multiplication process. The final outputs are available from the register file from
the P_x, P_y registers.

Finite field inversion using the extended Euclidean algorithm
Input: A polynomial a, irreducible polynomial p of the field $GF(2^m)$
Output: b = $a^{-1}mod(p)$
1. $u \leftarrow a, v \leftarrow p$

1. $u \leftarrow a, v \leftarrow p$
2. $g_1 \leftarrow 1, g_2 \leftarrow 0$
3. While $(u \neq 1 \; and \; v \neq 1)$ do
    3.1. if $(u[0] = 0)$ then,
        3.1.1.1. $u \leftarrow (u \gg 1)$
        3.1.1.2. If $(g_1[0] = 0)$, then $g_1 \leftarrow (g_1 \gg 1)$; else
                $g_1 \leftarrow (g_1 + p) \gg 1$
    3.2. If $(v[0] = 0)$ then,
        3.2.1.1. $v \leftarrow (v \gg 1)$
        3.2.1.2. If $(g_2[0] = 0)$, then $g_2 \leftarrow (g_2 \gg 1)$; else
                $g_2 \leftarrow (g_2 + p) \gg 1$
    3.3. If $u > v,$ then $u \leftarrow u + v, g_1 \leftarrow g_1 + g_2;$
    else $v \leftarrow v + u, g_2 \leftarrow g_2 + g_1$
4. If $u = 1, return(g_1);$ else $return(g_2)$

Fig. 4 Extended Euclidean algorithm for field inversion

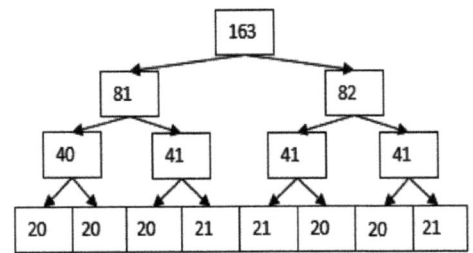

Fig. 5 Tree for 163-bit Karatsuba multiplier

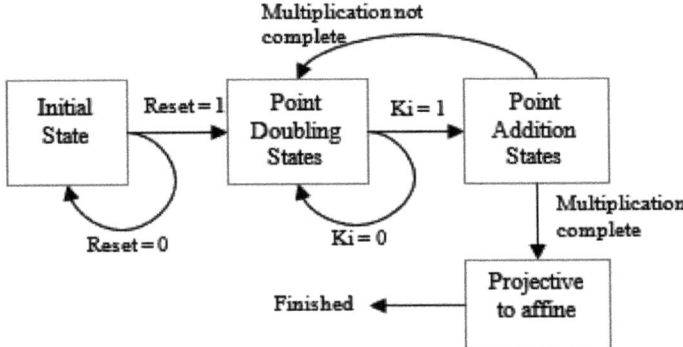

Fig. 6 FSM for PM control logic

From the point doubling and point addition algorithms, these operations will require a fixed number of states and hence a fixed number of clock cycles for execution.

Another aspect of this architecture is that the key register is placed separately from the register file. This physical separation of the key wherein it cannot be accessed by any unauthorized means introduces the hardware-level protection for the key. This physical boundary between the key and any malicious entity is one of the most important advantages offered by hardware-based implementations over conventional software implementations of cryptographic algorithms.

3 Discussions

The proposed design was implemented using the binary field $GF(2^{163})$ with the standard NIST recommended irreducible polynomial $p = x^{163} + x^7 + x^6 + x^3 + 1$. The curve equation was taken to be the NIST recommended binary curve $E: y^2+xy = x^3 + x + 1$, i.e. $a = b = 1$. The proposed architecture makes use of a 163-bit hybrid Karatsuba multiplier for field multiplications and dedicated inverter modules for field inversions. While this will increase the speed of execution, the associated trade-off in the area usage can be seen directly. The key register was 32-bit long. The comparisons of the proposed architecture with other published papers using the same FPGA and binary field $GF(2^{163})$ are given in Table 1. The number of clock cycles and the time are the values obtained for one point multiplication.

From the table, it can be easily seen that area-efficient implementations are much slower than the proposed design. However, the proposed design suffers from a very low maximum clock frequency which makes the time required for one point multiplication comparable to the other implementations. But implementations with an area usage which is comparable to the proposed design too require more time for the execution of one point multiplication.

The area of the proposed design is large due to the use of the dedicated inverter modules and the Karatsuba multiplier. The resource usage of these modules is given in Table 2.

The area usage indicates that using a more area-efficient multiplier can reduce the overall area requirement, at the cost of increased latency. The use of an inversion algorithm such as the Itoh-Tsujii algorithm can further reduce area usage as this

Table 1 Comparison of proposed design with other published works

Ref.	Slices	LUTs	Clock cycles	F_{max} (MHz)	Time (μs)	Field size	FPGA
[16]	16,209	26,364	3,010	154	19.55	163	Virtex-4
[17]	12,834	22,815	3,372	196	17.2	163	Virtex-4
[5]	3,536	6,672	4,168	290	14.39	163	Virtex-4
[18]	–	27,889	2,128	133	16	163	Virtex-4
[8]	4,080	7,719	4,050	197	20.56	163	Virtex-4
Ours	14,255	27,111	918	71	12.8	163	Virtex-4

Table 2 Resource usage of various modules in the proposed design

Module	Slices	LUTs	FPGA
Hybrid 163-bit Karatsuba multiplier	6,304	1,0982	Virtex-4
Field inverter	1,365	2,581	Virtex-4
Double squarer	181	315	Virtex-4
Squarer	95	165	Virtex-4
Adder	94	163	Virtex-4

algorithm uses the existing squaring and addition modules. The other field arithmetic modules, namely the adder, squarer and double squarer, have a negligible resource usage as compared to the multiplier and inverter.

4 Conclusions

This paper presents a fast implementation of the scalar point multiplier required for elliptic curve cryptography. The area usage, while large as compared to other similar implementations, is a justified trade-off for the speed the design possesses. The number of clock cycles required for the completion of one point multiplication is the lowest and thus shows the high-speed nature of the design, which can be used for many high-performance applications which do not have tight area constraints. The resource usage can be lowered by using other inversion algorithms and multipliers, but these optimizations will cause an increase in latency. The architecture for an entire elliptic curve crypto-processor using the elliptic curve Diffie–Hellman protocol can be constructed around this scalar point multiplier. A viable implementation of such a processor on an FPGA can use the embedded MPU found on most modern FPGAs such as the Artix-7.

References

1. Institute of Electrical and Electronics Engineers (2000) P1363 standard specifications for public key cryptography, NY
2. American National Standards Institute (1999) X 9.62 public key cryptography for the financial services industry: elliptic curve digital signature algorithm (ECDSA), Washington
3. National Institute of Standards and Technology (1994) FIPS 186-digital signature standard, Gaithersburg
4. Dormale GM, Quisquater J-J (2007) High speed hardware implementations of elliptic curve cryptography, a survey. Elsevier J Syst Archit 53:72–84
5. Khan ZUA, Benaissa M (2015) Throughput/area efficient ECC processor using montgomery point multiplication on FPGA. IEEE Trans Circuits Syst II Express Briefs 62(11):1078–1082

6. Khan ZUA, Benaissa M (2016) High-speed and low-latency ECC processor implementation over GF(2m) on FPGA. IEEE Trans Very Large Scale Integr (VLSI) Syst 25(1):165–176
7. Liu Z, Liu D (2016) An efficient and flexible hardware implementation of the dual-field elliptic curve cryptographic processor. IEEE Trans Ind Electron 64(3):2353–2362
8. Ansari B, Hasan MA (2008) High-performance architecture of elliptic curve scalar multiplication. IEEE Trans Comput 57(11):1443–1453
9. Roy S, Rebeiro C, Mukhopadhyay D (2013) Theoretical modeling of elliptic curve scalar multiplier on LUT-based FPGAs for area and speed. IEEE Trans VLSI Syst 21(5):901–909
10. Choi HM, Hong CP, Kim CH (2008) High performance elliptic curve cryptographic processor over GF(2163). In: 4th IEEE international symposium on electronic design, test and applications
11. Mahdizadeh H, Masoumi M (2013) Novel architecture for efficient FPGA implementation of elliptic curve cryptographic processor over GF(2163). IEEE Trans VLSI Syst 21(12):2330–2333
12. Hankerson D, Menezes A, Vanstone S (2004) Guide to elliptic curve cryptography. Springer
13. Anoop MS (2007) Elliptic curve cryptography: an implementation guide
14. Rebeiro C, Mukhopadhyay D (2007) Hybrid masked Karatsuba multiplier for 233(2) GF. In: 11th IEEE VLSI design and test symposium, Kolkata
15. Kodali RK, Amanchi CN, Kumar S, Boppana L (2014) FPGA implementation of Itoh-Tsujii inversion algorithm. In: IEEE international conference on recent advances and innovations in engineering, Jaipur
16. Chelton WN, Benaissa M (2008) Fast elliptic curve cryptography on FPGA. IEEE Trans VLSI Syst 16(2):198–205
17. Azarderakhsh R, Reyhani-Masoleh A (2012) Efficient FPGA implementations of point multiplication on binary edwards and generalized Hessian curves using Gaussian normal basis. IEEE Trans VLSI Syst 20(8):1453–1467
18. Fournaris AP, Zafeirakis J, Koufopavlou O (2014) Designing and evaluating high speed elliptic curve point multipliers. In: Euromicro conference on digital systems design

A Bloom Filter-Based Data Deduplication for Big Data

Shrayasi Podder and S. Mukherjee

Abstract Big data is growing at an unprecedented rate with text data having a large share and redundancy is a technique to ensure availability of this data. Large growth of unstructured text data hinders the primary purpose of the big data rendering the data difficult to store and search. Data compression is a solution to optimize the use of the storage space for big data. Deduplication is the most useful compression techniques. This paper proposes a two-phase data deduplication mechanism for text data. In the syntactic phase, a combination of clustering and Bloom Filter is used. In the semantic phase, a combination of SVD and WordNet synset is employed. Experimental results show the efficacy of the proposed system.

Keywords Deduplication · Bloom Filter · Clustering · SVD · WordNet

1 Introduction

A large amount of data to the tune of terrabytes are generated every day by various mediums. Eighty percent of generated data is claimed to be unstructured and in the text format [1], making data management a difficult, time-consuming, and complex task. The big data is primarily characterized by the rate of growth of the data. This data is stored in various storage mediums including cloud. To contain such a massive amount of data, storage continues to grow at an explosive rate (52% per year) [2, 3]. By the end of 2020, the size of the total generated data will surpass 30 zettabytes (ZB) as per a very conservative estimation [4, 5].

However, generating new data is just one quantum of the real problem of the growth of the data. Unprecedented growth of the data is also contributed to other,

S. Podder
Institute of Engineering and Management, Kolkata, India
e-mail: pshrayasi@gmail.com

S. Mukherjee (✉)
DIST, CEG, Anna University, Chennai, India
e-mail: msaswati@auist.net

© Springer Nature Singapore Pte Ltd. 2018
M. L. Kolhe et al. (eds.), *Advances in Data and Information Sciences*, Lecture Notes in Networks and Systems 38, https://doi.org/10.1007/978-981-10-8360-0_15

more complex, data handling mechanisms employed. To handle data easily, data needs to be at the location of the user, and in the digital world today, a user may be in any geographic location. While with the growth of the Internet, it is possible to serve the data to a user anywhere, bandwidth and other issues pose as serious challenges. To resolve this problem, typically data would be duplicated and distributed to various geographic locations for ease of access, thereby relieving the need of bandwidth usage and networking across large geographical regions. However, this causes a unique problem of redundant duplicated data occupying more and more storage space. Redundancy in the big data is contributing proactively to the rapid increase in the size of this data. Although new forms of storage such as Cloud employ different techniques to improve storage efficiency, it is a costly method, especially in the scenario where the demand of a specific data reduces after an initial period. To alleviate this problem, such duplicates must be managed, which entails ensuring that the unnecessary duplicates are detected and deleted; a challenging task, given the nature and size of the big data.

Typically, the technique of deduplication is widely used to identify and remove similar contents from various data storages, thereby reducing cost and saving space in data centers [6]. Deduplication is a very effective method that claims to save up to 90–95% of storage. Even though data deduplication technique had evolved as a simple storage optimization technique for secondary storages, it is widely adapted in larger storage areas like cloud storage. Different data deduplication is widely used by various cloud storage providers like Dropbox [7], Google Drive [8]. Deduplication is becoming a popular method for not only in backups and archives, but also in primary data centers. The major benefit of deduplication is saving disk space, and it also reduces the search space for searching such data.

While applying deduplication mechanisms is difficult in numeric data, it is particularly challenging in case of text data since, besides having identical data blocks, text data may also be similar in the semantic sense. Besides identifying and removing identical data, an efficient deduplication method for text data must identify and eliminate semantically similar data as well.

To handle deduplication of text data effectively, this research proposes a two-stage mechanism that can efficiently deduplicate a large amount of text data. The first phase is the syntactic similarity phase, where two documents are subjected to a process of detecting similarity by comparing the set of words in the documents. A combination of clustering and a proposed Bloom Filter called Updatable Bloom Filter is used in this phase. This mechanism can identify identical documents. Semantic similarity is handled in the second phase. This is the process of detecting semantic similarity between two documents by not only comparing the set of words in the documents but also comparing their synonyms. A combination of SVD and WordNet synset is used in this phase.

The paper is organized as follows. Section 2 investigates the existing work in the area of deduplication, clustering, and Bloom Filter. Section 3 discusses the proposed method. In Sect. 4, algorithms are described. Section 4 shows the efficacy of the system using the experimental results with Sect. 5 bringing the conclusion.

2 Background Details and Related Work

Over the years, many efficient solutions to the deduplication problem by measuring document similarity and by using Bloom Filter were proposed. Su et al. [9] suggested the method of using an integer Bloom Filter in cloud storage. The algorithm used for hashing in Bloom Filter is Secured Hash Algorithm (SHA-1). A number of hash functions have been defined. Data is divided into chunks that are processed in different clusters and are checked for duplicates using Bloom Filter in each cluster. As a result, unique data are written on the storage. Merlo et al. [10] suggested a method of deep learning, called Self-Organizing Maps (SOM). da Cruz Nassif et al. [11] used clustering algorithms like k-means and k-medoids. K-medoid is the process of taking median of data points in a cluster. K-means is the process of taking mean of data points in a cluster. After this step, hierarchical clustering techniques like centroid-based, average-link based were used to identify similar documents. Combinations of k-means and k-medoid with hierarchical clustering techniques were performed and the one with higher accuracy was chosen. Jiang et al. [12] suggested a simple method of using a feature to check if the documents are similar. Feature can be a word or a set of words. Feature can either appear in both documents, in any one document or, in none of the documents. Based on the number of features matched, similarity score is assigned between two documents. Pires et al. [13] proposed a vector space model which involved machine learning techniques namely k-nearest neighbors, random forests, and support vector machines. These techniques use cosine similarity as the distance measure. Semantic analysis has also been performed which included word co-occurrence, matching noun phrases, WordNet synonyms, etc. Gemmel et al. [14] implemented an idea of removing duplicate entries from a spreadsheet of names and addresses. The authors used affine gap distance, which is a variation of Hamming distance to check whether two records are similar. Historically, most deduplication-related publications focus on a narrow range of topics: maximizing deduplication ratios and read/write performance. Further, existing research work on document deduplication is either syntactic or semantic according to the need of the application. Machine learning techniques used are largely time-consuming and many used mechanisms cannot process very large documents. Attempts to modify clustering techniques to improve time have been performed. The proposed model aims at creating a new model for deduplication using machine learning and semantic techniques, attempting to optimize the limitations found by measuring the similarity of the data both syntactically and semantically.

3 Proposed Approach

In the proposed method, we employ a two-phase deduplication. The first phase performs deduplication based on the syntactic similarity, and the second phase is based on semantic similarity.

3.1 Syntactic Phase

To perform deduplication using syntactical similarity, it is required to find all pairs of documents that are identical to each other and replace one with a pointer to the other. This process can be recursively applied to all the documents for complete deduplication of a set of documents. However, in a big data environment, there will be a large number of documents in a data center and the method of applying such mechanisms consumes time as well as CPU cycles. This research proposes to optimize using two approaches. First, the documents are clustered such that each cluster contains lesser number of documents. Since clustering groups similar documents together, similarity need not be measured between documents across clusters. To create the initial clusters, we apply a k-means clustering. *K-means* clustering is dependent on the initial seed and its efficacy depends on the robustness of initial seed selection. To improve this method, we employ fractionation algorithm for obtaining the initial k seeds. The corpus is broken into n/m buckets each of size $m > k$. An agglomerative algorithm is applied to each of these buckets to reduce them by a factor of v. Thus, at the end of the phase, we have a total of $v \cdot n$ agglomerated points. The process is repeated by treating each of these agglomerated points as an individual record. This is achieved by selecting one of the documents within an agglomerated cluster. The approach terminates when a total of k seeds remain. These seeds are used as centroids in standard k-means clustering algorithm in order to determine good clusters.

A document is added to the cluster having the smallest cosine similarity value between the document and the corresponding centroid of that cluster. For every iteration, cosine similarity is calculated between the documents and centroids to decide whether a document should be moved to another cluster or not. The document which lies closest to the mean value of all the documents in a cluster is made the new centroid of that cluster. The above processes are repeated until the centroids do not change.

During the formation of the clusters, deduplication is performed by using proposed Forgetful Updatable Bloom Filter (FUBF). A Bloom Filter (BF) is a space-efficient probabilistic data structure that can be used to test whether an element is a member of a set.

A basic BF uses an array of m bits. The whole array is initially set to 0. When a document is to be inserted, it is hashed using a fixed number of hash functions, k. Therefore, the BF can be represented using parameters (m, k). Each hash output of a BF maps to a bit in the filter, which is then set to 1. To check for the presence of a document, the same set of hash functions are used to verify if all the mapped bits are already 1. In a Bloom Filter, it is possible to get false positives (i.e., a document not present may appear to be present) but no false negatives are possible (i.e., it always returns the right answer for a document not present) [15]. However, research shows that there have been ways to keep the rate of false positives to an acceptable level (almost close to 0) by manipulating parameters [16, 17]. Hence, Bloom Filters can be employed to check for the presence of a duplicate in a cluster during the process of creating the cluster, thereby avoiding altogether the incidence of adding an identical duplicate to a cluster.

While BF is an efficient way, it gives rise to two following problems that lead to scalability issue [18].

1. Nothing is forgotten by the BF. Hence, the bits set by older operations cannot be deleted.
2. Parameters (m, k) need to be fixed at the time of creation of the Bloom Filter.

As a result, once the Bloom Filter fills up, it is hard to "scale up" or "move" its elements to another BF with different parameters. Hence, a BF provides more and more false positives as it fills up since more bits are set to 1 by the incoming documents.

We propose to use a Forgetful Updatable Bloom Filter (FUBF) that employs the techniques of Forgetful Bloom Filter [18]. The proposed FUBF uses three filters for the representation of the documents in each cluster:

1. A current documents Bloom Filter;
2. A old documents Bloom Filter;
3. A new documents Bloom Filter.

All these Bloom Filters are equal in size and identical in their use of hash functions. Before a new document is to be inserted in a cluster, it is checked for the presence of its duplicate in the cluster using FUBF. All the three filters of the FUBF of the cluster are checked for this purpose. If the tested document is present in at least one of the three constituent filters, a duplicate is found and the new document is not inserted. However, if it is not found in any of the three filters, it is inserted only into the new and current Bloom Filters, but not in the old Bloom Filter. This way, the FUBF purges the filters of the older documents giving way to add more and more documents in a cluster, ensuring scalability.

At the end of the complete syntactic similarity phase, we obtain k clusters of documents having no syntactically similar documents. We perform semantic similarity on these clusters.

3.2 Semantic Phase

In the semantic phase, semantic similarities between documents are considered. We propose to use a variant of Latent Semantic Analysis (LSA) to check the semantic similarity. Documents in each cluster from syntactic phase are further divided into concept-based clusters using LSA that employs Singular Value Decomposition (SVD) method. We propose to apply LSA to obtain those documents that contain the same concept and calculate similarity between all pairs of such documents. If the similarity measure for a pair of documents is greater than a pre-defined threshold, these documents are subjected to the next level of semantic checking where WordNet synset is used for obtaining sentence level concept-based similarity. In this level, each document from a pair is split into sentences. Nouns and named entities are used to compare a sentence with sentences belonging to the other document in the pair. If the targeted words of two sentences have the same tag, then the sentences are considered to be similar. If all the sentences are similar, then the documents are considered as duplicates.

4 Experimental Setup and Results

The proposed method is tested against four different datasets as follows: a set of 6,000 documents which has details about different countries with 2,000 duplicates; a set of 10,000 documents having jokes that has 3,000 duplicates randomly injected and a set of 25,000 documents having jokes that has 10,000 exact and semantic duplicates; a set of 10,000 books from Gutenberg Web site with 2,000 exact duplicates. The proposed method is tested against two existing approaches. The first is constraint-based k-means clustering algorithm which is

Table 1 Deduplication using the proposed model

Dataset name	No. of data	No. of duplicates present	No. of duplicates found	F-measure
Country dataset	6,000	2,000	1,809	0.94
Joke dataset	10,000	3,000	2,521	0.91
Joke dataset	25,000	10,000	9,107	0.95
Gutenberg books	10,000	2,000	1,823	0.94

Table 2 Deduplication using constrained k-means

Dataset name	No. of data	No. of duplicates present	No. of duplicates found	F-measure
Country dataset	6,000	2,000	1,409	0.70
Joke dataset	10,000	3,000	1,921	0.66
Joke dataset	25,000	10,000	7,897	0.78
Gutenberg books	10,000	2,000	1,389	0.76

Table 3 Deduplication using chunking

Dataset name	No. of data	No. of duplicates present	No. of duplicates found	F-measure
Country dataset	6,000	2,000	1,609	0.87
Joke dataset	10,000	3,000	2,121	0.76
Joke dataset	25,000	10,000	8,970	0.88
Gutenberg books	10,000	2,000	1,598	0.84

Fig. 1 Improved results of the proposed system

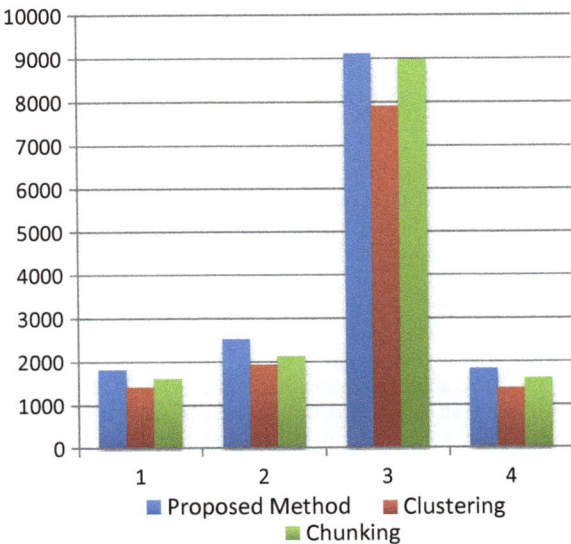

a deduplication method proposed by Hu et al. [19]. The second is deduplication using chunking method as suggested in [20]. These algorithms are applied on the same datasets and the corresponding results are compared. Table 1 shows the result of applying our proposed two-phase method on all four datasets. Table 2 shows the results of constrained k-means clustering method on all the datasets using 200 constraints. Table 3 shows the results of applying chunking for deduplication over the same datasets. F-measures are calculated for each method. Figure 1 shows the performance comparison of the three methods. It could be noted that the F-measure of the proposed model is higher than the other methods. It is evident that the extra levels of Bloom Filter during clustering and LSA and WordNet-based methods after clustering have increased the recall of the process, thereby improving F-measure.

5 Conclusions

From the results, it is clear that the performance of the proposed two-phase deduplication method is better than the pure clustering or chunking-based approach. The proposed method can be further improved by implementing concept-based clustering for the semantic phase.

References

1. CWADN, http://www.computerweekly.com/
2. Eaton C, Deroos D, Deutsch T, Lapis G, Zikopoulos P (2012) Understanding big data. McGraw-Hill Companies
3. https://www.smartfile.com/blog/the-future-forecast-for-cloud-storage-in-2018/
4. https://www.cisco.com/c/en/us/solutions/collateral/service-provider/visual-networking-index-vni/vni-hyperconnectivity-wp.html
5. Reed DA, Gannon DB, Larus JR (2012) Imagining the future: thoughts on computing. Computer 45
6. Deduplication, http://en.wikipedia.org/wiki/Data_deduplication
7. https://www.dropbox.com/
8. https://www.google.com/drive/
9. Su YH, Chuan HM, Wang SC, Yan KQ, Chen BW (2014) Quality of service enhancement by using an integer bloom filter based data deduplication mechanism in the cloud storage environment. In: IFIP international conference on network and parallel computing. Springer, Berlin, pp 587–590
10. Su YH, Merlo P, Henderson J, Schneider G, Wehrli E (2013) Learning document similarity using natural language processing. Linguistik Online 17(5)
11. da Cruz Nassif LF, Hruschka ER (2013) Document clustering for forensic analysis: an approach for improving computer inspection. IEEE Trans Inf Forensics Secur 8:46–54
12. Jiang J-Y, Lin Y-S, Lee S-J (2014) A similarity measure for text classification and clustering. IEEE Trans Knowl Data Eng 26:1575–1590
13. Pires CE, Nascimento DC, Mestre (2016) Applying machine learning techniques for scaling out data quality algorithms in cloud computing environments. Appl Intell 45:530
14. Gemmell J, Rubinstein BIP, Chandra AK. Improving entity resolution with global constraints. https://arxiv.org/abs/1108.6016
15. Bose P, Guo H, Kranakis E, Maheshwari A, Morin P, Morrison J, Smid M, Tang Y (2008) On the false-positive rate of bloom filters. Inf Process Lett 108(4):210–213
16. Bloom BH (1970) Space/time trade-offs in hash coding with allowable errors. Commun ACM 13(7):422–426
17. Wikipedia (2015) Bloom filter. https://en.wikipedia.org/wiki/Bloom_filter
18. Subramanyam R (2016) Idempotent distributed counters using a forgetful bloom filter. Clust Comput 19(2):879–892
19. Hu G, Zhou S, Guan J, Hu X (2008) Towards effective document clustering: a constrained K-means based approach. Inf Process Manag 44:1397–1409
20. Tolic A, Brodnik A (2015) Deduplication in unstructured-data storage systems. Elektroteh Vestn 82(5):233

ITDA: Cube-Less Architecture for Effective Multidimensional Data Analysis

Prarthana A. Deshkar and Parag S. Deshpande

Abstract Recent developments in real-time applications, sensor technology, and various online services are responsible for generating large amount of data which can be used for analysis. Performing multidimensional data analysis on such type of data requires aggregation at various levels which is generally done using data cubes. Generation of data cubes involves lot of storage and time overheads which make such approach practically less feasible if aggregation involves lot of hierarchies in dimensions. The Integrated Tool for Data Analysis (ITDA) project aims to provide a data analytics solution, under single Web-based platform to address the issue of generating the cube for high volume data by proposing the 'on-the-fly aggregation' architecture. This paper presents the ITDA which aims to provide the support for absorption of data, modeling it in multidimensional model, analyzing the absorbed data, and producing effective visualization. Target users can do analysis on their data without relying on costly tools or any prior knowledge in programming. In this paper, detailed architecture of ITDA software with its operating mode is discussed.

Keywords Multidimensional data analysis · Data mining · Cube
Cube-less architecture

P. A. Deshkar (✉)
Department of Computer Technology, Yeshwantrao Chavan College of Engineering, Nagpur, India
e-mail: Prarthana.deshkar@gmail.com

P. S. Deshpande
CSE Dept., G. H. Raisoni College of Engineering, Nagpur, India
e-mail: psdeshpande@cse.vnit.ac.in

P. S. Deshpande
Computer Science and Engineering, Visvesvaraya National Institute of Technology, Nagpur, India

© Springer Nature Singapore Pte Ltd. 2018   169
M. L. Kolhe et al. (eds.), *Advances in Data and Information Sciences*, Lecture Notes in Networks and Systems 38, https://doi.org/10.1007/978-981-10-8360-0_16

1 Introduction

In recent years, multidimensional analytics tools are becoming guide for data researchers as these tools give them cutting edge over their counterparts in marketplace. Due to increased frequency of data generation, data under consideration of analysis is also increasing tremendously. The large size of the data and complexity in data analysis demands an easy platform so that researchers and target users can do analysis on their data without the hard-core knowledge of information technology.

Ad hoc querying or ad hoc reporting is the main need of data analysis. To achieve this, data modeling is essential task if the system wants to facilitate the variety of domains. Multidimensional data modeling is the way to provide facility to perform ad hoc analysis. Analyzing multidimensional data is of growing need to extract the knowledge and hence to enable the decision making in various domains. Data analysis process, which leads to the enhanced decision making, combines various techniques like statistical techniques, data mining algorithms, and machine learning techniques. With all these techniques, presentation of analysis output with attractive visuals is a key part of popular analytics systems. Most of the current multidimensional systems rely on data cubes which are very much resource- and time-intensive. In this context, ITDA architecture is proposed to give the solution for multidimensional analysis with the reduced memory and time overheads as compared to the existing systems. The proposed system is providing analysis without the generation of cube.

Section 2 of this paper explains the preliminary concepts which are used in this paper. Section 3 provides the review of the systems which are already available. Section 4 proposes the new architecture. Section 5 elaborates the operating mode of the proposed architecture. Section 6 discusses the data security and privacy measures taken by the proposed architecture, and the last section is the conclusion of the proposed work.

2 Preliminaries

Multidimensional data is a type of data which talk about the fact which is associated with various entities called dimensions. Measure carries information of operational values such as sales, quantity. It represents the value of parameter on the basis of which multidimensional analysis is performed. Dimensions are informational entities such as product, region, and time, which are used to analyze the data. Generally dimensions have hierarchical structure. Dimensions may have additional properties like sequential relationship or more complex relationship.

For example, region dimension may have hierarchical relationship such as country, state, region, city. Time dimension has hierarchical as well as sequential properties. It can have days, weeks, months, and year as hierarchical levels. Sequential property indicates that each month is having value of the previous month and the next month. These relationships in the data play an important role in the analysis. The hierarchical

Table 1 Sample x values

Dimension	#Unique	Symbol	Value
Product	2	$X1$	CD, PEN DRIVE
Region	1	$X2$	PUNE
Year	2	$X3$	2015, 2016

relationship allows comparing fact values at different levels of hierarchy and can be used to design the market strategy for advertising requirements, supply chain management, to find out market share and so on. Sequential relationship helps to keep track of comparative progress among dimensions and within dimensions. For example, ratio of sales with previous month sales gives us growth over last month. Dependency relationship helps in the future prediction. For example, economists might base their predictions of the annual gross domestic product (GDP) on the final consumption spending within the economy. They use a dependency between GDP and final consumption spending.

Multidimensional data model is having multiple dimensions, and while analyzing it often the data at higher level of dimension is required; e.g., data generated at the seconds or hours level and for analysis data may be required at month, quarter, or year level. Multidimensional data analysis often requires such aggregated data, and hence these aggregations are stored in the form of cube to have faster data retrieval. A cube is used to generate aggregation of multiple dimensions or multiple combinations of multiple dimensions. As the number of rows and columns in the base data table having dimensions and facts increases, the cube generation may suffer from combinatorial explosion. Suppose there are d dimensions each with N unique rows, then the size of cube is given by N_d.

Let $x1, x2, x3, \ldots$ be the number of unique entries in each column except the measure column. In general, the number of rows in cube will be $(x1 + x2 + x3 + \ldots + xn) + (x1x2 + x2x3 + \ldots + xnx1) + (x1x2x3 + x2x3x4 + \ldots + xnx1x2) + \ldots + (x1x2 \ldots) + 1$. The extra addition of 1 is to include the final aggregation of fact (grand total). The following example represents the same with sample data (Table 1).

In the above example, number of rows in the cube can be calculated as:

$$(x1 + x2 + x3) + (x1x2 + x2x3 + x1x3) + (x1x2x1) + 1 = (2 + 1 + 2)$$
$$+ (2 * 1 + 1 * 2 + 2 * 2) + (2 * 1 * 2) + 1 = 18.$$

Need of Cube-Less Architecture

To perform the analysis in multidimensional environment, cube architecture is used to store the aggregates. Aggregated values facilitate the analysis process and reduce the response time of the system. But forming a multidimensional model for high-speed streaming data and generating data cube for analyzing it is a big challenge from the perspective of space and performance.

Let us take an example table containing seconds data to analyze the feasibility. The table under consideration would contain, say, 3 columns only. One is time column where data is recorded by a sensor which gives out values every second, and the second is the value/reading given by the sensor corresponding to each second. Third might be modes for those of which we are interested to see the value in each second. Even if we consider only 30 days data, it would result in table containing 5,184,000 rows. Further, running cube query on this table for a single aggregation would result in

$$2 * 2,592,000 + 25,920,002 = 6,718,469,184,000$$
$$= 6.72 * 1012, \text{rows}.$$

Hence, using cube architecture one has to wait for months to do some basic processing on such a high granular data. This is one aspect in the generation of cube. One more aspect is all these calculations are based on consideration that we are storing aggregations for only one aggregation function. But in real-life problems, analysts may require aggregations on the multiple aggregation functions like average, minimum, maximum. One more challenge in cube generation is relationship between the dimensions. Consider a scenario where two levels of a hierarchy, product and state, are present in the table. It is possible that multiple products are manufactured in a single state and multiple states manufacture a single product. So there is a many-to-many relationship between product and state. Cube generation has to be done twice, one considering a hierarchy of state under product and the other with a hierarchy of product under state. Again, it is inefficient in both space and time.

The proposed system is the research effort to overcome big challenge in handling the multidimensional stream data while generating the cube. It aims to develop a system which will address the storage overhead of cube architecture for stream data from any domain and will try to incorporate maximum statistical and data mining and machine learning algorithms for analysis. So the main objective of this system is to provide a single platform for multidimensional reporting, statistical processing, data mining, machine learning and visualization. The proposed system is designed on the basis of cube-less architecture, where aggregations are performed at query level and calculated on-the-fly, hence trying to reduce the time and storage overheads which are the main side effects of the cube architecture. The system is targeted expert as well as non-expert data miners.

3   Related Work

The proposed system aims to provide a data analytics, or decision support system works on the huge volume of data which comes from any domain. Many existing commercial products and research projects are working to facilitate the data analysts but with different approaches. Due to advancement in the technology, data generation speed is increasing exponentially and the communities want the knowledge from this data in very less time so that they can strengthen the decision support system.

Usman AHMED (2013) in the paper [1] proposed the technique to reduce the time required to get the analysis result from large amount of data, rightly said as 'analysis latency is pre-aggregation of data in a cube'. But cubing of data gives rise to problem like complexity in calculations and storage of data. In the same paper, to handle real-time data loading and thus avoiding the storage overhead of cube, they have proposed one approach to create blank tables with the same structure as that of source tables, then data is copied to it, then data is loaded in the data warehouse, and those temporary tables are removed.

B. Janet, A. V. Reddy (2011) in the paper [2] proposed the approach to manage or overcome the storage space issue. The approach is to use the subset of materialized view.

Authors Konstantinos Morfonios, Yannis Ioannidis (2006) in the paper [3] proposes another approach is by avoiding storage of unwanted and redundant aggregations. They have proposed a ROLAP cubing method called Cubing Using a ROLAP Engine (CURE) that computes whole data cube over very large data space constituted of hierarchical dimensions. CURE uses an efficient algorithm for partitioning fact table that helps improving the cube computation speed. One more approach is performing cube construction process in parallel [4].

Sandro Fiore, Alessandro D'Anca, Donatello Elia, Cosimo Palazzo, Ian Foster, Dean Williams, Giovanni Aloisio (2014) in the paper [5, 6] propose the decision support system for big data. Even if the systems are targeting big data as their data source, they are following the traditional OLAP structure to store the multidimensional data, i.e., cube architecture to store the data. Also some programming knowledge is required to customize the analysis, and hence system can focus on very limited set of data mining algorithms.

Many decision support systems are very much problem specific. Hence, the architecture and set of analytical algorithms are specific to that domain only, for example, analysis of text data [4], analyzing the stream data of clinical domain for classification. Zhang et al. [7] proposed 'on-the-fly' cube generation for sensors data [8], for analysis of the traffic data [9].

IBM research team (2008), in document [10], provides a dynamic cube architecture for a very popular commercial analytical system IBM Cognos. It is also creating a cube, using in-memory caching to support large database. They have implemented the approach by storing the once retrieved data from cube in caches, and if required retrieve from caches.

From the above discussion, it seems that the decision support system which handles the storage overhead of cube architecture and not restricted to particular domain is the need. Also it is observed that the systems which are developed carry very specific analytical features, or may require the expert knowledge in the data analysis.

This proposed architecture is addressing the storage overhead by avoiding the generation of cube, and the aggregations are done at query level or on-the-fly. Further optimization of the query structure is also planned to enhance the performance of the query execution. The proposed system is also offering the complete analytical processing including multidimensional reporting, statistical processing, data mining, machine learning, and visualization under single platform. Also the focus of system is not restricted to any specific domain or problem.

4 Proposed Architecture

The proposed system is basically designed to facilitate the researchers and data analyst with the complete package of multidimensional reporting, statistical processing, data mining, machine learning and visualization. This is achieved by the Web-based system with user-friendly and secure environment for the data analyst. Proposed system is functionally independent; this means it does not require any additional external component or system to complete the task. Also, the components of this proposed system are integrated and there is no need to install any of the components separately, which is often common for most of analytics tools.

Proposed system architecture is mainly divided into two parts, data modeling part and data analysis part. As the system is modeled as a Web application, user can access it as a client and the processing part is handled by the server. Client will act as a data provider, and the processing algorithms are residing at the server side. The main objective of the proposed cube-less architecture is to reduce the time and storage overhead which are the side effects of the cube architecture; hence, data storage on server is avoided (Fig. 1).

Proposed system consists of two main parts containing various components. First is data absorption from different data sources, collection of metadata, and formation

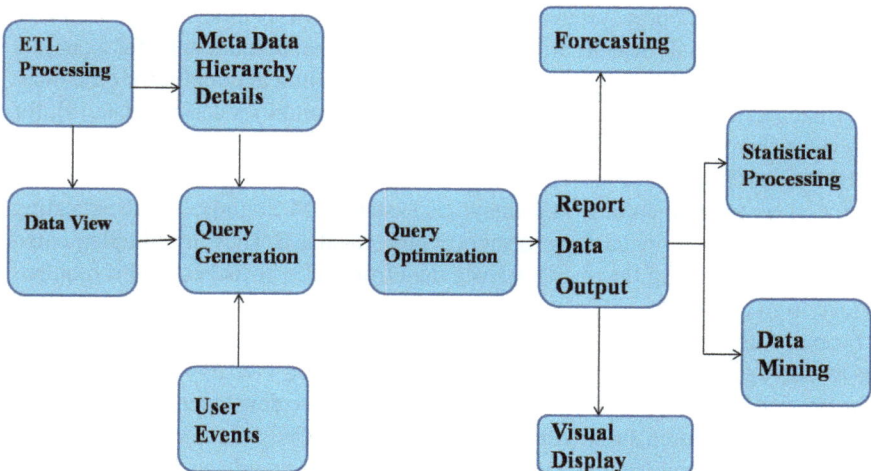

Fig. 1 Cube-less architecture

of multidimensional model, and second is multidimensional analysis on modeled data which further extends to perform statistical analysis and data mining.

Data modeling functionality mainly includes the extraction, transformation, and loading (ETL) process. Source data is given to the ETL process, and it produces the ready to analyze data. ETL process is responsible to extract the data residing on various sources and in variety of formats. It also performs cleansing and customization of data according to the analysis needs. This process is also responsible to generate the metadata of the ready to analyze data. The proposed system is not going to store the data and the aggregations; hence metadata is having crucial role in this system. Aggregations can be generated on-the-fly by using the metadata.

User events are nothing but the data requirements of the analyst to perform the MDDM. Target users of the system can be expert or non-expert data miners, and hence system provides the graphical user-friendly interface to select the data items. System will convert the user selection into user events. With the help of these user events and metadata, system will automatically generate the query. Queries are generated on-the-fly even for the aggregated data. Calculation of aggregated values is done at the runtime. And hence time and storage overhead are avoided by the system.

One query is generated to achieve the value of one cell in the multidimensional output. Though the system achieves the reduction in storage and time overhead, queries are generated in high number. To optimize the number of queries generated, system will further introduce the query optimization engine to have reduced number of queries.

Multidimensional data analysis report is generated by the system. This report is in grid form as well as it provides the attractive visualization techniques to represent the multidimensional output. This output can further be analyzed using different statistical techniques, data mining algorithms, forecasting, and machine learning algorithms. All these components are integrated into the system itself. All these components also provide the reports with grid values and the visual representation.

5 Operating Mode

Expert or non-expert data miners can access the system by creating client users. User will upload the data through these clients. The system is able to extract the data from flat files or from databases. For flat files system supports excel, csv, and txt file formats; for databases it supports MS-Access and Oracle formats. Using the graphical user-friendly interface, user can perform transformations on data to achieve the customization according to organization needs for analysis. The ETL process ends by generating the metadata of the final customized data prepared after transformation. During this entire process, one log file is maintained to record the activity log of user. It contains all status messages and stack trace in case of some exceptions and errors. During each step of execution of ETL algorithm, logged user first records its status and then performs the step. If there are some bad rows in the given dataset, then the process takes care of putting it to bad file which is also maintained by the ETL process. Along with these files, metadata information is also stored in the file, which is then used for further analysis. Metadata contains description of the hierarchy of the dimension, formats followed by the data items, database credentials, and customized formulas created by user to analyze the data. While deciding metadata parameters time dimension is handled separately. Time dimension requires the special treatment as it can have sequential as well as normal dimension properties. If it is normal time dimension, then its hierarchy and bounds for uppermost levels can be directly considered; i.e., if uppermost level for time is year, then we need to record lower and upper bound for year so that we can give interactive interface to user for time dimension data selection during analysis. If it is sequential dimension, then it is needed to have lower and upper bound for each level. In the terminology of ITDA, the proposed architecture, all these files are stored in the 'environment' of the client user. Environment is the conceptual area given to user along with the multidimensional model for the single data view. One user may have any number of environments for the absorption of data. Environment is in the form of folder on the client machine. Data is accessed with the help of environment. Because of the environment structure, user can handle the many-to-many relationships present in the dimension, separately. At the time of implementation, the metadata information which is stored in file is passed to server in a form having smaller grammar and easy to map with the data structures used by the modern programming language. Only transfer of metadata instead of complete data will reduce the transfer and storage overhead.

After creation of the environment, now system is made ready to perform the analysis. To generate the multidimensional output, user needs to provide the data which is to be analyzed. In ITDA terms, these user requirements for analysis are termed as user events. The proposed system provides different mathematical facilities to prepare the data for analysis. Users are allowed to specify the aggregation function on which data aggregation is required. User is free to select more than one aggregation function at a time as the aggregations are not going to be stored in the system, so no need to worry about time and storage required for the aggregations. Along with the aggregation functions, the system provides various filters to filter the data which is to be analyzed.

Rank Filter Rank filter provides an easy way to add inputs based on the rank of input values.

Measure Filter Measure filter is similar to rank filter. Here, the user would want to select inputs based on the measure value's range instead of rank value.

In the multidimensional report, along with the aggregated values for dimension values, user also can have some analytical functions for the preliminary analysis. The analytical functions which are provided in the system are dense rank, cumulative distribution, ntile, percent rank to one, market share, growth rate, etc.

With the help of all the information provided through the user events, query builder engine generates the queries automatically. Query builder module takes the input selections for each of the dimensions in row and column sequences and builds a single query for each combination.

NR_{di} be the number of selection from dimension i present in row sequence.
$NCdi$ be the number of selection from dimension i present in column sequence.

$$\mathtt{Total\ number\ of\ queries\ fired = p\ i\ NRdi\ X\ p\ i\ NCdi}$$

The query complexity can be given by $N \times k$ where N is number of rows in the table and k is number of cells in the output matrix.

Once the row and column sequences are specified, the outputs from the execution of queries are stored in a $R \times C$ matrix where R is product of number of inputs for all dimension specified in row sequence and C is the product of number of inputs for all dimension specified in column sequence.

There is a specific mapping from output matrix to the graph being plotted. The measure is plotted along the y-axis. Each one of the rows in the matrix is mapped to x-axis. The columns in matrix are mapped as legends which create overlapping plot.

6 Security and Privacy

The proposed architecture is maintaining the security and privacy by giving different privileges to users according to their need. As the proposed system is going to store only the metadata for the environments created by user on the client side, no other user can access the metadata. To maintain the privacy, the system is managed by the authorized login, i.e., admin login. Admin is responsible to allow any user to create their account. No other user can access the data from the other user.

To use this system, user first needs to create the account. According to level of understanding and experience in the data analysis domain, there are four privileges given to user while creating the account. If the user is expert data miner, then the user is having right to upload the data in the system, can create the multidimensional data model by its own, and then can perform multidimensional analysis on it. Professional analysts having data but wants to avoid the technicality of creation of multidimensional model, then they can just upload their data, and admin is going to create the environment for them and notify them after successful creation of environment. Some users can have only right to perform analysis using statistical and data mining algorithms on the environments created and assigned by the admin. Here in this case, user is not having access to source data. And for the non-expert users, there is facility that admin will assign some predefined reports to study the analysis.

Every user can create the environments based on the analysis need. N number of environments can be created for a single dataset. These environments are stored on the client side, hence not accessible to any other user.

7 Conclusion and Future Work

In this work we have presented the system ITDA, a complete solution for data analysis. This model follows the cube-less architecture to overcome the side effects of cube architecture like storage requirement and time to build a cube. In this paper, we talk about the architecture of the system and its components. Also we focus on the operating mode of the system so that system capabilities can be explored. We propose a system which can perform multidimensional reporting, statistical processing, data mining, machine learning and visualization under single platform. In future, the query optimization part of the system needs to be developed so that time required to generate on-the-fly queries can be optimized.

References

1. Ahmed U (2013) Dynamic cubing for hierarchical multidimensional data space. Ph.D. thesis
2. Janet B, Reddy AV (2011) Cube index for unstructured text analysis and mining. In: ICCCS'11, 12–14 Feb 2011, Rourkela, Odisha, India
3. Morfonios K, Ioannidis Y (2006) CURE for cubes: cubing using a ROLAP engine. In: VLDB'06, 12–15 Sept 2006, Seoul, Korea
4. Jin D, Tsuji T (2011) Parallel data cube construction based on an extendible multidimensional array. In: 2011 International Joint Conference of IEEE TrustCom-11
5. Fiore S, D'Anca A, Elia D, Palazzo C, Foster I, Williams D, Aloisio G (2014) Ophidia: a full software stack for scientific data analytics. 978-1-4799-5313-4/14/$31.00 ©2014 IEEE
6. Fiore S, D'Anca A, Palazzo C, Foster I, Williams DN, Aloisio G (2013) Ophidia: toward big data analytics for eScience. In: 2013 international conference on computational science. https://doi.org/10.1016/j.procs.2013.05.409
7. Zhang Y, Fong S, Fiaidhi J, Mohammed S (2012) Real-time clinical decision support system with data stream mining. J Biomed Biotechnol
8. Mehdi M, Sahay R, Derguech W, Curry E (2013) On-the-fly generation of multidimensional data cubes for web of things. IDEAS'13 09–11 Oct 2013, Barcelona, Spain
9. Geisler S, Quix C, Schiffer S, Jarke M (2011) An evaluation framework for traffic information systems based on data streams. Elsevier Ltd. All rights reserved
10. IBM Cognos Dynamic Cubes, Oct 2012

On Using Priority Inheritance-Based Distributed Static Two-Phase Locking Protocol

Sarvesh Pandey and Udai Shanker

Abstract Two-phase locking with high priority (2PL-HP), a well-suited concurrency control protocol for distributed real-time database systems (DRTDBS) because of being free from priority inversion problem, is used for accessing data items to resolve conflicts among the concurrently executing transactions. However, it suffers from the problems of wastage of system resources responsible for degrading the system performance. In DRTDBS, our basic aim is to minimize the number of transactions missing their deadline. In this paper, static two-phase locking with priority inheritance (S2PL-PI) protocol has been proposed specifically to minimize the wasted system resources, i.e., CPU and data items by avoiding unnecessary abort of transactions by optimal use of priority inheritance mechanism. A DRTDBS is simulated for comparison of the performance of S2PL-PI protocol with previous other protocols, and results confirm the significant improvement in system performance.

Keywords Concurrency control · Two-phase locking · 2PL-HP · Priority inheritance · Distributed real-time database

1 Introduction

Database systems (DBS) are a collection of logically interrelated data items shared by multiple users [1, 2]. A user can interact with databases by means of a partially ordered set of read and write actions termed as a transaction. Every transaction follows ACID (Atomicity, Consistency, Isolation, and Durability) property [3, 4]. Isolation is one of the essential transaction properties. The role of isolation comes into play when more than one transactions concurrently execute in the database using

S. Pandey (✉) · U. Shanker
Department of Computer Science and Engineering, MMM University of Technology,
Gorakhpur, UP, India
e-mail: pandeysarvesh100@gmail.com

U. Shanker
e-mail: udaigkp@gmail.com

© Springer Nature Singapore Pte Ltd. 2018
M. L. Kolhe et al. (eds.), *Advances in Data and Information Sciences*, Lecture Notes in Networks and Systems 38, https://doi.org/10.1007/978-981-10-8360-0_17

critical resources [5]. To ensure the isolation among concurrently executing transactions, the system must govern the interaction among them by having a serializable schedule. Concurrency control (CC) schemes ensure the serializability among concurrently executing transactions [6]. Results obtained from concurrently executing transactions can be reflected in the database only when there exists an equivalent serial execution schedule of such concurrently executing transactions. CC schemes are broadly classified as pessimistic and optimistic. In pessimistic CC scheme, conflict is detected before access of data item, while in optimistic CC scheme, conflict is detected after the access of data item. After detection of conflict among a set of concurrently executing transactions, a conflict resolution mechanism comes into play. Conflict resolution mechanism specifically does the following.

1. Select transaction(s) from the set of conflicting transactions to prosecute.
2. Take an appropriate action against selected transaction(s) at a suitable time.

The two most used actions are blocking and abort. If a conflict is detected before data item access, one of both actions can be taken. However, abort action is appropriate in case of conflict detected after data item access. Several CC schemes have been proposed to overcome the problem of inconsistency [6, 7]. Most of the techniques used to facilitate concurrent execution of transactions rely on the notion of locking of data items. The commercial conventional DBS uses two-phase locking (2PL) [8] as a CC scheme. They use locking of data items to ensure isolation among concurrently executing transactions and thereby guaranteeing serializability. Every data item in a DRTDBS is associated with a variable that describes the possible operations that can be performed on it. 2PL protocol says that all locking operations on a transaction precede the first unlock operation. In 2PL, every transaction obtains a lock before accessing any data item. Transaction execution divides into two phases: growing phase and shrinking phase [9]. A transaction can acquire a lock on data items during growing phase, but any of the locks that are acquired during this phase cannot be released. In shrinking phase, a transaction can release all the locks acquired during growing phase, but cannot acquire any new lock.

Two widely known 2PL variants available in the literature are static 2PL and dynamic 2PL. In static 2PL, the data item access list, i.e., locks required by a transaction are presumed to be granted prior to the start of its execution [10]. The transaction locks all needed data items before it begins its execution. Here, a data item access list is predeclared before execution of any transaction. Note that a data item access list consists of a set of data items required to be granted access for completion of a transaction. If any of the predeclared data items of the data item access list cannot be granted an access, then the transaction does not lock any of the data items; however, it waits until all data items are available for locking. In dynamic 2PL, transactions obtain locks to access data items on request and release locks upon expiry or commit. The working principle of static 2PL is like dynamic 2PL except the technique of setting locks on data items. In general, static 2PL requires smaller number of messages for setting locks as compared to dynamic 2PL. Static 2PL does not suffer from deadlock since blocked transactions cannot hold a lock on any data item.

In non-real-time static 2PL, a transaction gets blocked if any of its required lock(s) from the predeclared data item access list is/are locked by some other transactions. At the time of transaction blocks, some of the data items required by it may be free. Such data items that were free at the time when conflict occurred can be seized by other transactions later. As a result, even after the original conflicting data items are released, the transaction may get blocked by other transactions arrived after it. Consequently, the requesting transaction blocking time can be randomly long because of extended blocking which is a result of waiting for more than one locks. Real-time S2PL (RT-S2PL) protocol [11] overcomes this problem. Here, each data item in the database is assigned a priority equal to the priority of highest priority transaction from the set of transactions that have requested access to this data item. Like S2PL, all the data items required to be accessed by a transaction need to be locked before starting execution of a transaction. If a conflict occurs for any of the data items, none of the required locks will be assigned to a requesting transaction. Note that in case a data item requested by a transaction has lower priority than that of the transaction itself, its priority will be updated to that of the requesting transaction. All such features of RT-S2PL lead to its suitability for DRTDBS.

The 2PL wait promote (2PL-WP) protocol [12, 13] is identical to 2PL in its resolution of conflicts. As in 2PL, 2PL-WP resolves the conflict (if any) by means of blocking a requesting transaction. It is the first concurrency control protocol based on priority inheritance mechanism to reduce the negative impact of priority inversion problem [14], which is inherent in a real-time environment. In case of priority inversion, low-priority lock-holding transaction inherits the priority of the highest priority transaction from the set of transaction requested access to the data item. Further, lock-holding transaction retains this inherited priority until it either commits or restarted. Priority inheritance mechanism specifically reduces the priority inversion duration so that requesting high-priority transactions can get the conflicting resource earlier [15, 16].

The 2PL high-priority (2PL-HP) protocol [12, 13] ensures that low-priority transactions do not delay high-priority transaction by eliminating priority inversion problem. It does so by resolving data conflicts instantly in favor of the transaction with higher priority. 2PL-HP concurrency control protocol suffers from the problem of the cyclic restart. Just like 2PL-HP, the S2PL high-priority (S2PL-HP) protocol also [1] does not suffer from priority inversion problem. It does so by resolving data conflicts instantly in favor of the transaction with the higher priority. Lengthy transactions suffer from the starvation problem due to use of S2PL-HP. One more severe problem with S2PL-HP protocol is that it may lead to undesirable wastage of resources due to ABORT of low-priority lock-holding transactions in case of conflict with high-priority transaction. This may lead to the increase in a number of transactions missing their deadline. S2PL-HP concurrency control protocol suffers from the problem of the cyclic restart. Let us consider an example to explain this problem.

Suppose, at time t_1, a distributed real-time transaction T_1 starts its execution at site S. Coordinator of transaction T_1 divides it into a set of subtransactions as $T_1 = \{T_{11}, T_{12}, T_{13}, ..., T_{1N}\}$, where N is the number of cohorts participating to complete T_1. All these subtransactions execute at different sites in the database system. At time

t_2, subtransaction T_{12} is in the processing phase and completed locking phase. The data item O is locked by T_{12}, during its locking phase. Note that execution period consists of locking phase and processing phase, respectively. This same data item O is further required by some other transaction T_2 that enters in the system. Note that priority of T_2 is greater than the priority of T_1. So, as per S2PL-HP, the transaction T_1 is aborted (or restarted), and the data item O will be given to T_2. Note that all the subtransactions participating to complete the execution of T_1 will be restarted, no matter conflict arrives only at the site where subtransaction T_{12} is running. There is a possibility that after a restart of transaction T_1, its priority may become higher than the priority of T_2 as a result of decrease in slack time, which directly affects the deadline of the transaction. In such case, transaction T_2 is aborted (or restarted) and transaction T_1 will again lock the data item O. This leads to the cyclic restart problem among transactions (transactions T_1 and T_2). Transactions involved in cyclic restart miss their deadline at the end, which in turn leads to the wastage of resources and degradation in performance of the system in terms of increase in a number of transactions missing their deadline.

Based on S2PL, S2PL-HP, and 2PL-WP, static two-phase locking with priority inheritance (S2PL-PI) protocol has been proposed which optimistically uses the priority inheritance mechanism to minimum resource wastage. It also improves a chance of the low-priority transaction to get completed even after the conflict with some high-priority transaction provided that high-priority transaction can manage to wait for completion of low-priority transaction. S2PL-PI also overcomes the starvation problem with lengthy transactions up to some extent.

Section 2 discusses proposed S2PL-PI concurrency control protocol in detail. The performance study presented in Sect. 3 shows significant performance improvement in S2PL-PI protocol over other protocols. Finally, in Sect. 4, conclusions are drawn with future directions of works.

2 S2PL-PI: A Real-Time Concurrency Control Protocol

The lifetime of a cohort is divided into two phases, i.e., execution phase and commit phase. In SWIFT protocol, execution phase is further divided into two phases, i.e., locking phase and processing phase [17]. During locking phase, the cohorts lock all the required data items, and then, during processing phase, the cohort does some necessary computation. WORKSTARTED message to the coordinator is sent before the start of processing phase. If there are dependencies, then the sending of WORK-DONE message is deferred till the removal of dependencies. S2PL-HP is used as a concurrency control protocol in SWIFT protocol which is a combination of S2PL and 2PL-HP. Further, it is suggested that a focused and coordinated research work is required to develop a new concurrency control protocol which directly affects the performance of SWIFT and all the existing commit protocols.

In case a high-priority distributed transaction T_H requests access to the data item that is already locked by some low-priority lock-holding transaction T_L, then conflict resolution strategy is used to resolve such conflict that affects the system's performance. Although it is clear that minimum the wastage of resources (CPU and data items) because of ABORT, maximum the chance of successful completion of transaction, but at the same time, there is need to ensure that the concurrency control algorithm is more focused toward respecting priority of transaction rather than minimizing wastage of resources or increasing throughput of the system. Hence, a new S2PL-PI protocol has been proposed which optimistically minimizes wastage of resources, reduces the starvation problem, and in turn minimizes the number of transactions missing their deadline. In S2PL-PI protocol, the ABORT of lock-holding cohort is done based on the intermediate priority of all the cohorts of a lock-holding transaction T_L and the priority of lock-requesting transaction T_H. It is a temporary priority assignment policy without affecting the initial priorities [18] and is based on the remaining execution time (T_{Remain}) needed by the lock-holding low-priority cohorts ($T_{L1}, T_{L2},..., T_{Li}$) and the slack time available with the newly arrived higher-priority cohort (T_H). It also solves the problem of starvation of long cohorts that arises due to the high probability of access conflicts. The slack time is the amount of time the distributed transaction can afford to wait in order to complete before its deadline. The remaining execution time of the lock-holding cohorts ($T_{L1}, T_{L2},..., T_{Li}$) is given as:

$$T_{Remain} = R_i - T_{Elapse}$$

where R_i is the minimum transaction response time; T_{Remain} is remaining execution time needed by T_L; T_{Elapse} is elapsed execution time of T_L.

There are three ways to minimize the number of transactions missing their deadline.

1. If T_L is in the execution phase and Max$T_{Remain}(T_{L1}, T_{L2},..., T_{Li})$ is less than the slack time (T_H), then the priority of T_L gets inherited to T_H, and T_H is inserted into the wait queue.
2. If T_L is in the execution phase, Max$T_{Remain}(T_{L1}, T_{L2},..., T_{Li})$ is greater than or equal to the slack time (T_H), and T_L, then the low-priority lock-holding transaction gets aborted.
3. If T_{Li} is in commit phase (have sent a PREPARED message to its coordinator), then the priority of T_L gets inherited to T_H, and T_H is inserted into wait queue, no matter the requesting transaction T_H is a high-priority transaction. Although, the priority inheritance applied here, gives the conflicting transactions a chance of successful completion by reducing the priority inversion duration.

In brief, S2PL-PI protocol optimistically minimizes the wastage of resources by using priority inheritance mechanism and overcomes the starvation problem with lengthy transactions up to some extent. The following algorithm shows how the locks are granted in S2PL-PI.

2.1 S2PL-PI Algorithm to Resolve Data Conflict

Input: T_H is a high priority transaction requesting access to the data item O.

T_L is a low priority transaction that has locked the data item O.

$T_{L1}, T_{L2}, ..., T_{Li}$ are the subtransactions of a transaction T_L

BEGIN

S2PL-PI_lock_acquire ()

{

for each Data item, di

if (! lockConflict)

      assign a data item to T_H;

  else

  {

check priority of conflicting cohorts of T_L holding the data item;

if (priority (T_H) > MaxPriority $(T_{L1}, T_{L2}, ..., T_{Li})$)

        {

if (conflicting cohort T_{Li} haven't sent PREPARED message to its

coordinator, and is in processing phase of execution period)

             {

              if (slack time(T_H) $\Box$MaxT$_{Remain}$ $(T_{L1}, T_{L2}, ..., T_{Li})$)

              //*slack time(T_H) ≥ 0 *//

                 {

                        insert T_H in wait queue;

                        T_L inherits priority of T_H;

                 }

              else

              {

                        T_H aborts T_{Li};

                        allocate data item to T_H;

              }

             }

        else

            {

                  insert T_H in wait queue;

                  T_L inherits priority of T_H;

            }

        }

else T_H waits for completion of T_L;

  }

}

END

2.2 Major Contributions

The major contributions of S2PL-PI protocol are as follows.

1. Minimization of number of Aborts optimistically by use of priority inheritance scheme, and intermediate priority assignment policy (based slack time
   of lock-requesting high-priority transaction and remaining execution time of
   lock-holding low-priority transaction), which in turn minimizes the number of
   transactions missing their deadline.
2. Overcomes the starvation problem with lengthy transactions up to some extent
   using the intermediate priority assignment policy. Such problem arises due to the
   high probability of access conflicts in case of lengthy transactions.

3 Performance Evaluation

In the DRTDBS research community, there is no hands-on benchmark available to
assess the performance of the proposed protocol. Therefore, a DRTDBS including N
sites was simulated in accordance with the environment assumed in earlier studies
[17, 19–21]. We ensured a significant level of resource and data contention during
performance study. Table 1 presents different parameters used in a simulation study
with their default values.

Parameter	Meaning	Default setting
DB_{Size}	Size of database (no. of pages in databases)	200 data objects/site
N_{db}	No. of database sites	4
AR	Transaction arrival rate per site	0–4 transactions/sec (uniformly distributed)
T_{com}	Communication delay among transactions	Either 1 ms or 100 ms
N_{op}	No. of operations in transaction	4–20 (uniformly distributed)
SF	Transaction slack factor	1–4 (uniformly distributed)
$P(w)$	Probability of write operation	0.60
CPU_{page}	Processing time required for accessing CPU page	5 ms
$Disk_{page}$	Processing time required for accessing disk page	20 ms

Earliest deadline first (EDF) is used as the cohort's priority assignment policy
for the performance study of S2PL-PI protocol. As per EDF, a transaction with the
closest deadline is assigned the highest priority in the system. In case of a tie, we

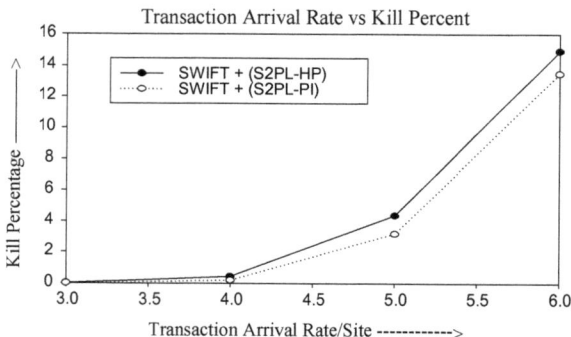

Fig. 1 Transaction kill percentage with resource and data contention at 1 ms communication delay under normal load

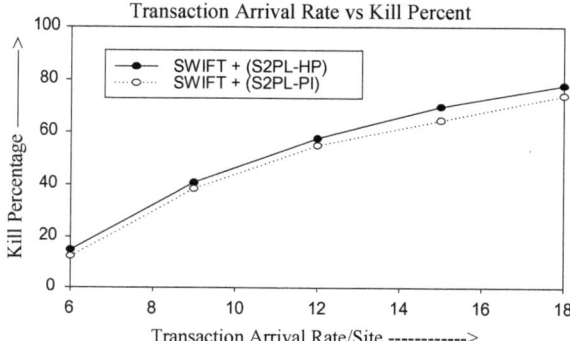

Fig. 2 Transaction kill percentage with resource and data contention at 1 ms communication delay under heavy load

assign priority to the transaction using FCFS scheme. The performance of S2PL-PI protocol is measured based on the number of transactions missing their deadline. Mathematical calculation of kill percent is done as the following,

$$\text{Kill Percent} = \frac{\text{Number of transactions aborted}}{\text{Total number of transactions in the system}}$$

3.1 Simulation Study and Performance Results

To investigate the performance of the S2PL-PI when applied to a DRTDBS, a wide range of operations ($N_{op} = 4$–20) in global as well as local transactions are introduced. The simulation study is performed with disk-resident databases. We compared the S2PL-PI concurrency control protocol with S2PL-HP protocol. Figures 1, 2, and 3 show the transaction kill percent at communication delay of either 1 ms or 100 ms in disk-resident databases with different transaction arrival rates.

The proposed protocol performs better than SWIFT protocol+S2PL-HP protocol under all load conditions. This is because of avoidance of ABORT of lock-holding transaction wherever possible by using priority inheritance scheme.

Fig. 3 Transaction kill percentage with resource and data contention at 100 ms communication delay under normal and heavy load

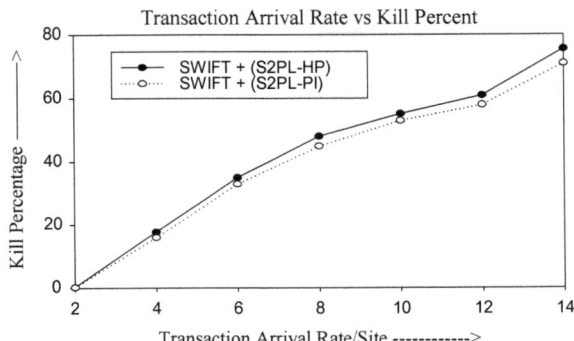

4 Conclusions

In this paper, the S2PL-PI protocol has been proposed to minimize the unnecessary abort of the transaction via the optimal use of priority inheritance mechanism. Here, an intermediate priority assignment policy has been introduced to assign an intermediate priority to the conflicting transactions at the time of data contention between them. This policy avoids the wastage of resources such as CPU and data item by not aborting a near to completion lock-holding transaction provided that a high-priority lock-requesting transaction can wait for the conflicting data item without missing its deadline or at least one of the lock-holding conflicting cohort is in PREPARED state. In this way, cooperative execution of conflicting transactions may lead to successful completion of all the competing transactions. It provided performance benefits over SWIFT protocol+S2PL-HP by optimum use of priority inheritance scheme and combination of initial and intermediate priority assignment strategies. A DRTDBS is simulated for the comparison of the performance of S2PL-PI protocol with previous other protocols, and results confirm the significant improvement in system performance.

As a part of future work, an exhaustive real-life implementation work is required to establish this approach as a value-based commercial product.

Acknowledgements We acknowledge the financial support provided by the Council of Scientific and Industrial Research (CSIR), New Delhi, India under grant no 1061461137 during this research work.

References

1. Shanker U, Misra M, Sarje AK (2008) Distributed real time database systems: background and literature review. Int J Distrib Parallel Databases 23(02):127–149
2. Shanker U, Misra M, Sarje AK (2001) Hard real-time distributed database systems: future directions. IIT Roorkee, India, pp 172–177

3. Pandey S, Shanker U (2016) Transaction execution in distributed real-time database systems. In: Proceedings of the international conference on innovations in information embedded and communication systems, pp 96–100
4. Ramamritham K (1993) Real-time databases. Distrib Parallel Databases 01(02):199–226
5. Faleiro JM, Abadi DJ (2015) FIT: a distributed database performance tradeoff. Data Eng 38(01):10–17
6. Yu PS, Wu K-L, Lin K-J, Son SH (1994) On real-time databases: concurrency control and scheduling. Proc IEEE 82(01):140–157
7. Kao B, Garcia-Molina H (1993) An overview of real-time database systems. Real Time Comput 127:261–282
8. Faleiro JM, Abadi DJ (2014) Rethinking serializable multiversion concurrency control. VLDB 08(11):1190–1201
9. Harding R, Aken DV, Pavlo A, Stonebraker M (2016) An evaluation of distributed concurrency control. VLDB 10(05):553–564
10. Lam KY (1994) Concurrency control in distributed real time database systems. Ph.D. thesis
11. Lam K-Y, Hung S-L, Son SH (1997) On using real-time static locking protocols for distributed real-time databases. Real-Time Syst 13(02):141–166
12. Abbott RK, Molina HG (1992) Scheduling real-time transactions: a performance evaluation. ACM Trans. Database Syst 17(03):513–560
13. Haritsa JR, Carey MJ, Livny M (1992) Data access scheduling in firm real-time database systems. Real-Time Syst 04(03):203–241
14. Pandey S, Shanker U (2018) A one phase priority inheritance commit protocol. In: Proceedings of the 14th international conference on distributed computing and information technology (ICDCIT), Bhubaneshwar, India, 11–13 Jan 2018 (Accepted)
15. Huang J, Stankovic JA, Towsley D (1991) On using priority inheritance in real-time databases. In: Real-time systems symposium, pp 210–221
16. Huang J, Stankovic JA, Ramamritham K, Towsley D, Purimetla B (1992) Priority inheritance in soft real-time databases. Real-Time Systems, vol 04, no 03, pp 243–278
17. Shanker U, Misra M, Sarje AK (2006) SWIFT—a new real time commit protocol. Distrib Parallel Databases 20(01):29–56
18. Shanker U, Misra M, Sarje AK (2005) Priority assignment heuristic to cohorts executing in parallel. In: 9th international conference on world scientific and engineering academy and society (WSEAS)
19. Lee VCS, Lam KW, Hung SL (2002) Concurrency control for mixed transactions in real-time databases. IEEE Trans Comput 51(07):821–834
20. Ulusoy O (1995) A study of two transaction-processing architectures for distributed real-time data base systems. J Syst Softw 31(02):97–108
21. Qin B, Liu Y (2003) High performance distributed real-time commit protocol. J Syst Softw 68(02):145–152

A New Way to Find Way Using Depth Direction A*

Akashdeep Singh, Ankit Agrawal, Pratik Patil, Priyanshu Pal and Prashant Udawant

Abstract The paper presents a review on a new algorithm for path finding, namely Depth Direction A* Algorithm, which uses linear graph theory in collaboration with A* Algorithm. The algorithm increases the efficiency of A* Algorithm for avoiding obstacles and finding the path of minimum distance to its goal in any map. Further, the path is more naturalistic, data structure requires less number of nodes for expansion and computation, and less memory is required. To prove this, we compare the results obtained by using various pathfinding algorithms such Dijkstra's, BFS, DFS, A*, and Depth Direction A*.

Keywords Pathfinding · Depth Direction A* · A* · RTS

1 Introduction

Pathfinding algorithms are used extensively in variety of fields of work and also in day-to-day life such navigation, computer simulation, artificial intelligence, traffic planning, and network design [1]. Path finding is especially used in the field of computer video games to move an agent (object or a character) from one position to

A. Singh · A. Agrawal · P. Patil · P. Pal · P. Udawant (✉)
Department of Computer Engineering and Information Technology, NMIMS/MPSTME,
Mumbai, India
e-mail: prashant.udawant@nmims.edu

A. Singh
e-mail: axedeepsin@gmail.com

A. Agrawal
e-mail: anki.agrawal24@gmail.com

P. Patil
e-mail: pratzpatil@gmail.com

P. Pal
e-mail: priyanshu2295@gmail.com

© Springer Nature Singapore Pte Ltd. 2018
M. L. Kolhe et al. (eds.), *Advances in Data and Information Sciences*, Lecture Notes
in Networks and Systems 38, https://doi.org/10.1007/978-981-10-8360-0_18

Fig. 1 Example of RTS game

another position avoiding obstacles such as water obstacles (sea plants, rocks, etc.), land obstacles (trees, rocks, mountains), and sometime air obstacles too [2].

In 3D space, the objects refer to 2D coordinate plane for maintaining their positions. Most of the strategy games [3] will play on terrain that is large and complex in relation to the movable agent, and there are many obstacles present in the environment. Pathfinding is an important element of RTS games [4]. The games involve many movable agents being moved in various paths at different times. The player does not have time to control every individual unit to proceed to where it needs to be at every instant [1, 4]. The artificial intelligence in the game is used to determine how an agent should proceed to a given goal node. This is done with the help of pathfinding algorithms. These algorithms seek for the most efficient path of minimum distance from the source point to the destination point (Fig. 1).

Besides, in RTS games, units and objects might have movement in multiple layers; that is, some agents can travel in water tiles, travel on ground, and fly above in the sky. Thus, each layer consists of an access cost value and the obstacles and barriers are different for different objects based on their ability to traverse the given layer. In other note, the game has rendering process that consumes time while rendering the game in 3D space [2, 5]. The experiment must show that rendering times are stable, so it does not affect the calculations regarding the pathfinding algorithms in independency.

2 Literature Survey

The analysis of path of minimum distance is an implementation of artificial intelligent which is developed with respect to the human capability of thinking, understanding the cause and effect, and learning [6]. Heuristic is defined as proceeding to a solution by trial and error or by rules that are only loosely defined. Heuristic technique is a technique for problem solving, learning, or discovery that employs a practical method which may not be optimal or perfect but is sufficient for immediate goals. It might return values that are not expected [1, 5]. This varies with each different problem. 3D RTS game uses analysis of minimum distance to manage the path of movement of movable agents. The research proposes the following idea. It proposes the comparison between several pathfinding algorithms in terms of rapidity, intelligence, memory consumption, and efficiency [3, 5]. The algorithms to be compared are breadth-first search, A*, depth-first search, iterative deepening, bidirectional breadth-first search, Dijkstra's, and depth-limited depth-first search [5]. The pathing analysis provides appropriate paths to move from start/source location to end/destination location for agents in variety of games. It can solve the trouble of finding the best possible shortest path while avoiding obstacles.

2.1 Breadth-First Search

Pseudo code:
```
bfs(Graph, start_node, end_node)
  {
     Create Empty List VISITED
     Create Empty  QUEUE
     VISITED.add(start_node)
     QUEUE.enqueue(start_node)
     current  =  start_node
     while current ! = end_node
     if v not in VISITED
        current  =  QUEUE.dequeue()
        for all neighbours n of current in Graph
        if n not in VISITED
           QUEUE.enqueue(n)
        end for
        VISITED.add(current)
     end if
     end while
  }
```

Scope for Improvements:
Breadth-first search [7] requires the saving of each level of traversal tree for further processing. Consequently, this leads increased memory usage and will easily exhaust the available memory in low-memory systems (Fig. 2).

Further, the time consumed by the algorithm is directly proportional to distance between the goal node and the root node.

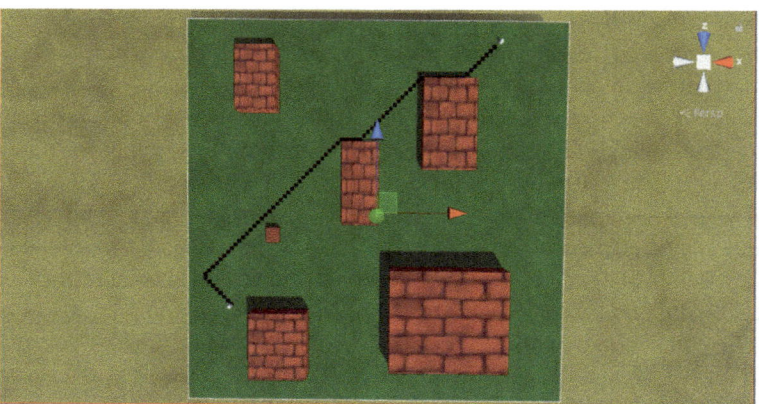

Fig. 2 Implementation of breadth-first search

2.2 Depth-First Search

Pseudo Code:
dfs(Graph, start_node, end_node)
{

        Create Empty List VISITED
         Create Empty STACK
        VISITED.add(start_node)
        STACK.pust(start_node)
        current = start_node
        while current ! = end_node
        current = STACK.top()
        STACK.pop()
        for all neighbours n of current in Graph
        if n not in visited
                STACK.push(n)
                VISITED.add(n)
        end if
        end for
        end while

}

Scope for Improvements:

Assuming infinite (very large value) depth of the search tree, then there is a possibility that the algorithm will continue to search along the left descendent node forever if in case the goal is not reached. To fix this, a depth cut-off value could be defined. This will further need to be optimized as the position of the goal node is not known; thus, if the goal node is below the cut-off, then goal will never be reached. Hence, it is not guaranteed to find the solution. Also, if loops exist in the graph,

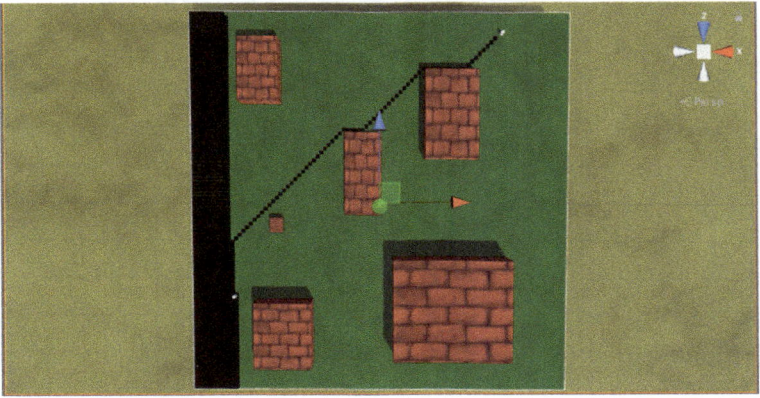

Fig. 3 Implementation of depth-first search

then DFS [8] will traverse these loops infinitely. For this, checking for loops must be implemented (Fig. 3).

2.3 Dijkstra's Algorithm

Pseudo Code:

```
dijkstra(Graph, start_node, end_node)
Create Empty List VISITED
Create Empty List NEIGHBOUR
current = start_node
VISITED.add(current)
while current ! = end_node
for each neighbour n of current in GRAPH
NEIGHBOUR.add(n)
end for
current = NEIGHBOUR.min()
VISITED.add(current)
end while
```

Limitations:

The time taken is a lot in Dijkstra's [6]. The path found may not always be the shortest as it does not take into account a heuristic function. The number of expansions made is very large, so a lot of memory is used (Fig. 4).

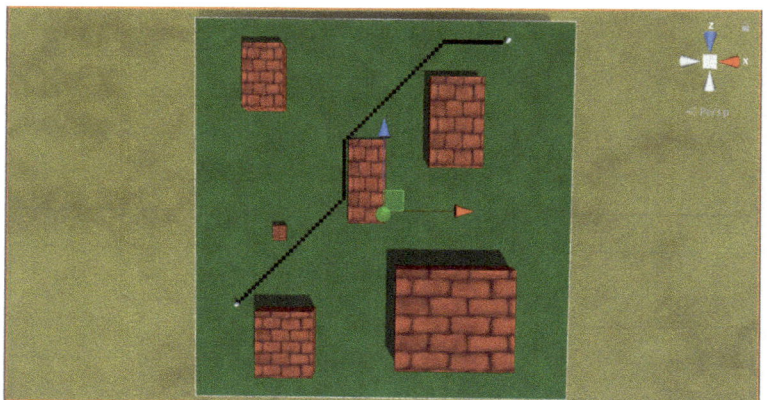

Fig. 4 Implementation of Dijkstra's

2.4 A*

Pseudo Code:
Astar(Graph,start_node,end_node)
{
        Create Empty List CLOSEDSET
        Create Empty List OPENSET
        OPENSET.add(start_node)
        while OPENSET.count > 0
        find the node in OPENSET with lowest f cost and call it current
        current.visited = true
        CLOSEDSET.add(current)
        OPENSET.remove(current)
        if current == end_node
                display path found
                return
        end if
        for all node n adjacent to current
        if not n.visited
                n.g_cost = current.g_cost + distance between g and n
                n.h_cost = distance between n and end_node
                n.f_cost = n.g_cost + n.h_cost
        end if
        if n not in OPENSET
                OPENSET.add(n)
        end if
end for
end while
}

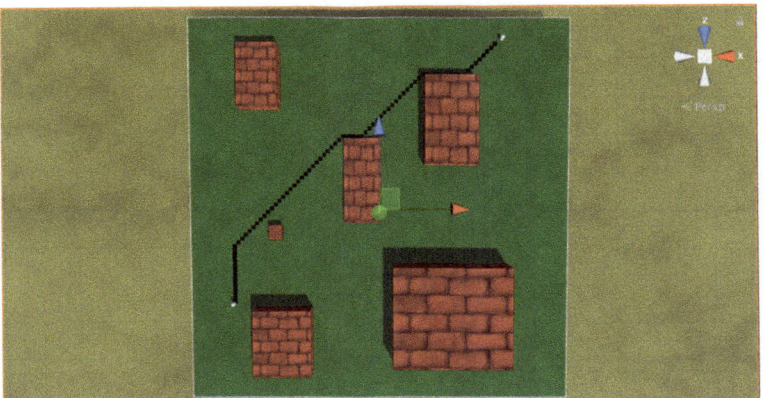

Fig. 5 Implementation of A*

Scope for Improvement:

The path that is formed does not look good. The number of nodes expanded in A* [4] is more compared to Depth Direction A* (Fig. 5).

2.5 Depth Direction A*

Depth Direction A* [5] is a modification of A* in which when no obstacles are found in the vicinity of the current node where the path has reached, it calculates the next node by using graph theory using the formula

$$y = mx + c \tag{1}$$

where y is the y index and x is the x index and m is the slope of line from the current point to the final node and c is the intercept of that same line. m is calculated using the formula

$$m = \frac{y2 - y1}{x2 - x1} \tag{2}$$

where $x2$ and $y2$ are the x and y coordinate of target node, respectively, and $x1$ and $y1$ are the x and y coordinate of the current nod. Similarly, c can be found by substituting value of x, y, and m in the equation of line. Thus, using m, c, and the new value of x coordinate, i.e., either old x coordinate $+1$ or old x coordinate -1, we can find the value y coordinate for the next node in the path, whereas if we encounter an obstacle we use the traditional A* Algorithm to avoid the obstacle.

We also make use of a heap instead of list for OPENSET, as we use a heap it will reduce the time required to find the element which has the lowest f cost and finding

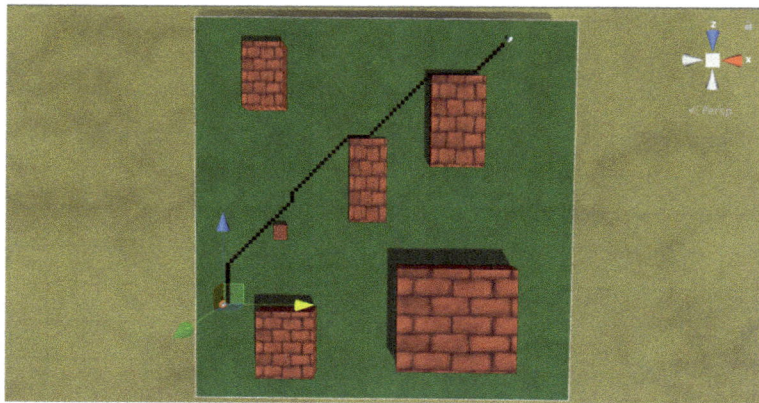

Fig. 6 Implementation of Depth Direction A*

the element which has the lowest f cost is a very costly affair where most of the time is spent by the algorithm (Fig. 6).

Depth Direction A* pseudo code:
Depth_Direction_Astar(Graph,root)

```
{
            Find intial value of m = (Goal Y – Start Y)/(Goal X – Start X)
            Create Empty List CLOSESET
            Create Empty Heap OPENSET
            OPENSET.add(root)
            While OPENSET not empty
            current = OPENSET.removefirst()
            if current's neighbors not obstacles
                        find next node using linear graph theory
            else
                        find next node using A*
            end if
            end while
}
```

3 Experiment

The algorithms were implemented using unity in which a separate class was used for each algorithm. We also had a class called node which stored was used to define a particular node in the map or grid. This class had a property called walkable which was a Boolean. This property was used to define if that particular node could be traversed by the object or if this node was to be considered an obstacle (Fig. 7).

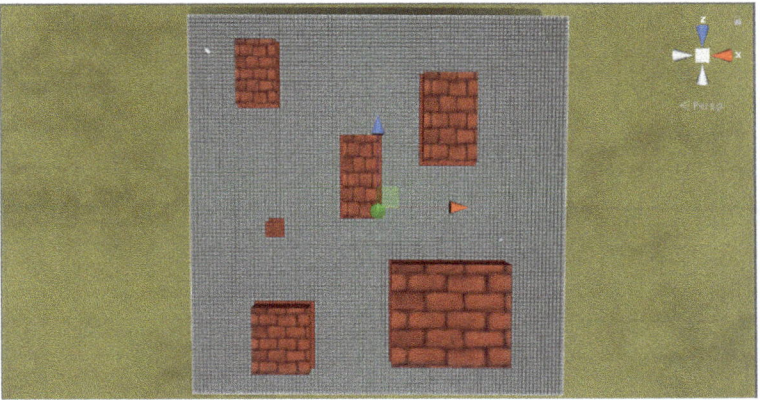

Fig. 7 White squares have property walkable as true, and red square has property walkable as false

Another class that was used was the grid class, which contained a 2D array of nodes for all the nodes in the map. The initialization of the walkable property of the node is done in this class. Since we also allow an object to have certain height before it become an obstacle, we check that in this class. The maximum allowable height is defined by the property maxHeight. This class also contains a function called GetWorldPointFromNode (Node A) which returns the x and y coordinate for node A.

We have implemented A*, Dijkstra, DDA* using a heap for instead of a list as this makes the searching of elements which have the least cost much less expensive. If we used list, we would have to search the whole list for the shortest element; even if we used a priority queue, we would have to traverse the queue a lot during insertion. This problem is handled by heap very effectively.

4   Result Analysis

The implementation of Depth Direction A* on unity sands the path developed by it to reach from one node to another node.

The comparison graph between Depth Direction A*, A*, BFS, DFS, and Dijkstra's is shown in the following graphs (Figs. 8 and 9).

The graphs represent time taken to reach the goal node (written above each graph) from start node by different algorithms. The comparison graphs show that Depth Direction A* pathing algorithm takes less time than any other pathing algorithm.

After deliberate comparison of aforementioned algorithms, we have come to a result that Depth Direction A* Algorithm is better in rapidity, efficiency, and reduced memory consumption.

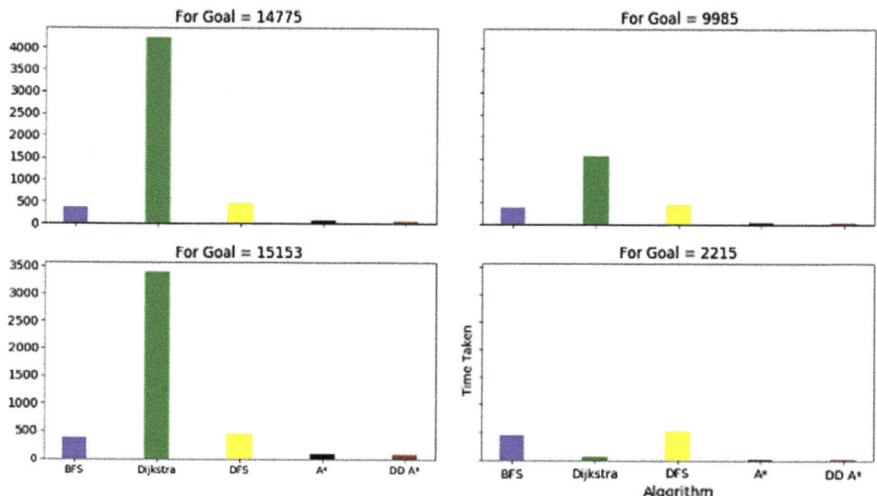

Fig. 8 Multiple comparisons for random map #1

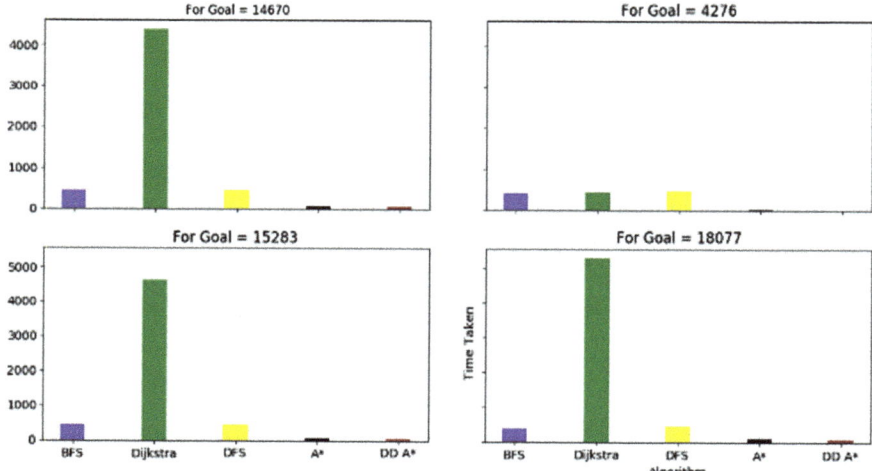

Fig. 9 Multiple comparisons for random map #2

5 Conclusion

In finality, after executing different pathfinding algorithms, including Depth Direction A* (DDA*), on various maps with varying obstacles, it was observed that DDA* Algorithm performed better in terms of space required and time taken. There was considerable amount of difference in time taken as well as the memory used for node expansion by DDA* for pathfinding. Further, the path observed in DDA* is neater, more linear, and naturalistic as compared to other pathing algorithms (Fig. 10).

Fig. 10 Green area is traversable, and red blocks are obstacles

The average time taken by A* is 9.87 ms, whereas the average time taken by DDA* is 6.07 ms. The difference is observed using a small map. The superior nature of DDA* over A* is observed more profoundly as the size of map increases. This is a comparison made when both A* and DDA* are implemented using heaps.

References

1. Educational Overview of AI (1997) The history of artificial intelligence
2. Hu J, Wan WG, Yu X (2012) A pathfinding algorithm in real-time strategy game based on unity3D
3. Padmanabhan V, Goud P, Pujari AK, Sethy H (2015) Learning in real-time strategy games. In: 2015 14th international conference on information technology
4. Ng PHF, Li YJ, Shiu SCK (2011) Unit formation planning in RTS game by using potential field and fuzzy integral. In: 2011 IEEE international conference on fuzzy systems
5. Khantanapoka K, Chinnasarn K (2009) Pathfinding of 2D & 3D game real-time strategy with depth direction A* algorithm for multi-layer. In: Eighth international symposium on natural language processing
6. Yang C-W, Lee T-H, Huang C-L, Hsu K-S (2016) Unity 3D production and environmental perception vehicle simulation platform. IEEE-ICAMSE
7. Angel D (2015) A breadth first search approach for minimum vertex cover of grid graphs. In: IEEE sponsored 9th international conference on intelligent systems and control (ISCO)
8. Tarjan R. "DEPTH-FIRST SEARCH .ANI) LmEAR GRAm ALGORITHMS

ECG Biometric Analysis Using Walsh–Hadamard Transform

Ranjeet Srivastva and Yogendra Narain Singh

Abstract The electrocardiogram (ECG) signal expresses unique cardiac features among individuals. This paper proposes a novel method to human identification using ECG. The proposed method utilizes a band-pass filter for quality check and autocorrelation (AC) for feature extraction. Furthermore, the Walsh–Hadamard transform (WHT) is used for feature transformation. To get cost- and time-efficient classification performance, the dimensionality of feature vector is reduced using linear discriminant analysis (LDA). Experimental results show the best identification rate of 95 and 97% over MIT-BIH arrhythmia database and QT database, respectively.

Keywords Human identification · Electrocardiogram · Walsh–Hadamard transform · Discriminant analysis

1 Introduction

The identity of a person needs to be determined in many applications of access control. Traditional identity verification methods based on passwords and ID cards are vulnerable to identity theft [1]. In order to offer better security to identity proving systems, many body parts and behaviors are being used from last decade [2, 3]. This class of strategy offers better security to identification system. Among them, some modalities are widely accepted but lack to provide robustness to circumvention or replay attacks and user privacy. In the recent years, researchers have suggested that physiological signals like electrocardiogram (ECG) have potential to be used for identity proofing and provide robustness in identification [4, 5].

R. Srivastva (✉)
Department of Information Technology, Babu Banarasi Das Northern India
Institute of Technology, Lucknow, Uttar Pradesh, India
e-mail: ranjeetbbdit@gmail.com

Y. N. Singh
Department of Computer Science and Engineering, Institute of Engineering
and Technology, Lucknow, Uttar Pradesh, India

© Springer Nature Singapore Pte Ltd. 2018
M. L. Kolhe et al. (eds.), *Advances in Data and Information Sciences*, Lecture Notes
in Networks and Systems 38, https://doi.org/10.1007/978-981-10-8360-0_19

In the literature, it has been shown that the ECG signals of different individuals are heterogeneous [6, 16]. The discriminatory features in ECG are found among individuals due to different levels of ionic potential, plasma level of electrolytes as well as physical structure, position, and size of heart. Different methods are found in the literature to analyze the ECG and its use to biometric application. One of the early studies of use of ECG biometric was presented by Biel et al. [6]. They have used multivariate method on a group of 20 subjects and achieved 100% identification rate. Shen et al. [7] have investigated the feasibility of ECG as a biometric by using time-domain and appearance-based features. They have achieved 95 and 80% classification accuracy by template matching and decision-based neural network approaches, respectively. By combining these two approaches, the reported identification result is 100% for 20 subjects.

Singh et al. [8–10] have analyzed the ECG using signal processing techniques. They have classified the individuals using a variety of features those are extracted from temporal, amplitude, and angle features with an accuracy of over 99%. Plataniotis et al. [11] have introduced a non-fiducial feature extraction method based on autocorrelation (AC) and discrete cosine transform (DCT). They reported the recognition rate of 100% on a data set of 14 healthy subjects. Chan et al. [12] have classified 50 subjects with accuracy of 95%, using three different quantitative measures: residual difference, correlation coefficient, and distance measure using wavelet transform.

A short-time frequency method has been developed by Odinaka et al. [13]. They have performed experiments on a sample of 269 subjects. The equal error rates of verification are found to be 5.58% on multisession data. When training and testing samples are collected from same day, the verification results are improved further. Agrafioti et al. [14] have presented an autocorrelation-based approach in conjunction with DCT and linear discriminant analysis (LDA). Wang et al. [15] have demonstrated the comparison of fiducial-based approach using analytic and appearance attributes and non-fiducial-based approach using AC and DCT. Li et al. [16] have proposed a hybrid approach fusing temporal and cepstral information and achieved identification accuracy of 98.26% on 18 subjects.

The issues related to these studies include individuality of ECG over larger population, sensitivity to exact localization of fiducial points, heart rate variations, different anxiety level. In this paper, a novel method is proposed that addresses the issues like the individuality of ECG and accurate localization of dominant fiducials. The method calculates the AC coefficients from the windows of filtered ECG signal. Further, the AC coefficients are transformed into WHT coefficients and LDA is applied to reduce the dimensionality. Experimental results show that the proposed method outperforms other methods on MIT-BIH arrhythmia database and QT database. The rest of the paper is outlined as follows: Sect. 2 presents the novel method of ECG analysis and its characterization that is used for biometric application. The experimental results are presented in Sect. 3. Finally, the conclusion is noted in Sect. 4.

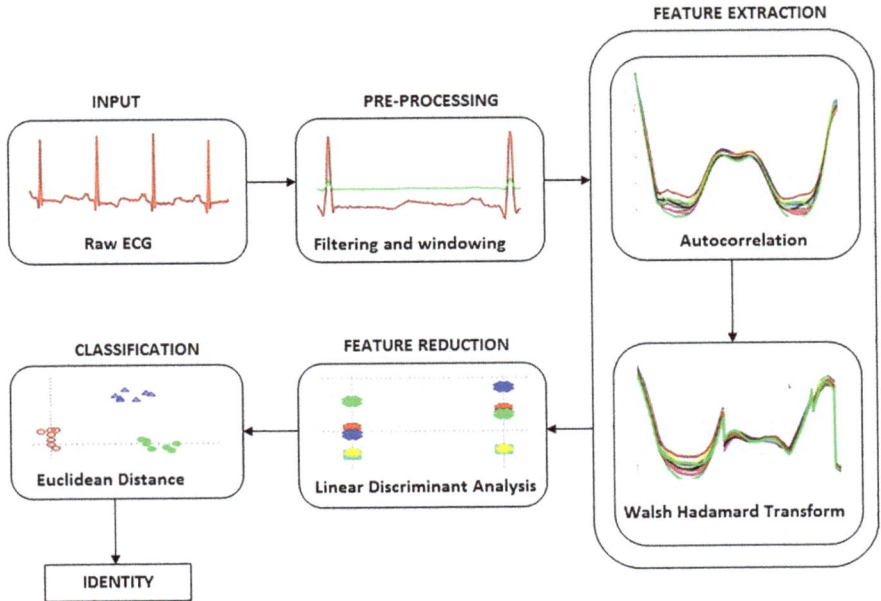

Fig. 1 A schematic of proposed ECG biometric system

2 Proposed Methodology

A schematic diagram of proposed biometric system is depicted in Fig. 1. It involves mainly: preprocessing, feature extraction, feature reduction, and classification. Different types of noise and artifacts are removed in preprocessing step. Features are extracted from an ECG trace of 50 seconds, by autocorrelation followed by Walsh–Hadamard transform (WHT). The LDA is used for feature reduction, and the last step of the identification process is classification based on match scores of the subjects.

ECG signals may have different type of noises such as low-frequency noise components including baseline oscillations, respiration or body movements and high-frequency noise components due to power line interferences. A band-pass filter is used to eliminate the effects of noise by combining a low-pass (Eq. 1) and a high-pass (Eq. 2) filter [17]. The cutoff frequency of low-pass filter and high-pass filter is about 11 and 5 Hz, respectively. The reason behind selecting this combination of filter is that most of the energy of ECG signal lies within the above frequency range.

$$y_n = 2y_{n-1} - y_{n-2} + x_n - 2x_{n-6} + x_{n-12} \tag{1}$$

$$y_n = 32x_{n-16} - (y_{n-1} + x_n - x_{n-32}) \tag{2}$$

The filtered ECG signal is divided into non-overlapping segments. The windowing criteria are followed with the motivation that the maximum correlation among data samples can be found, if the window size is at least two heartbeats. The length of window can be chosen heuristically according to the sampling rate of signals. For this experiment, all the data is sampled at 200 Hz, and the data window of size 50 seconds is chosen.

The fiducial-based feature extraction techniques may not achieve better classification performance, since it is highly dependent on the accurate localization of dominant fiducials of ECG wave. Several factors may affect the exact delineation of fiducial points such as noise present in the ECG signal. This motivates us to adopt a method which is independent of fiducial points of ECG signal. To extract features from ECG signal without localization of fiducial points, autocorrelation (AC) is applied on windowed ECG. The AC shows similarity of samples as a function of time lag between them. The AC provides an automatic, shift-invariant representation of similarity features over multiple cardiac cycles. The normalized AC ($\widehat{AC}_{yy}[t]$) for ECG signal, $y[i]$ of length n can be computed as follows,

$$\widehat{AC}_{yy}[t] = \sum_{i=0}^{n-|t|-1} \frac{y[i]y[i+t]}{\widehat{AC}_{yy}[0]} \tag{3}$$

where $y[i+t]$ is determined by shifting windowed ECG with a time lag of $t = 0, 1, \ldots (m-1)$; $m << n$.

The AC coefficients are transformed using WHT to maximize the inter-class dissimilarity and intra-class similarity. Walsh function offers a fast method of solving nonlinear differential and integral equations with the reduction in calculation speed and storage space [18]. The Walsh function has only three possible values:+1 or −1 in the interval $0 \leqslant x \leqslant 1$ and a value zero outside this interval. The Walsh transform of a given series of numbers $x_0, x_1, x_2 \ldots x_{N-1}$ can be calculated as follows,

$$a_j = \frac{1}{N} \sum_{t=0}^{N-1} x_t * w_j(x_t), \quad j=0,1,\ldots N\text{-}1 \tag{4}$$

where N is the number of samples in the series, and w_j is the Walsh function calculated as:

$$w_j(x) = 0, \quad for \quad x < 0 \, or \, x > 1 \tag{5}$$

$$w_0(x) = 1, \quad for \quad 0 <= x <= 1 \tag{6}$$

$$w_{2j}(x) = w_j(2x) + (-1)^j w_j[2(x-1/2)] \tag{7}$$

$$w_{2j+1}(x) = w_j(2x) - (-1)^j w_j[2(x-1/2)], \quad for \quad j=0,1,\ldots N\text{-}1 \tag{8}$$

The feature vectors formed with Walsh coefficients have higher dimension. To retain the discriminatory information even with lower dimension, LDA is applied to the feature vector. The LDA seeks to reduce dimensionality while preserving as much the class discriminability as possible. It linearly transforms the feature characteristics in a lower dimension space. More formally, let us assume that training set $\chi = \{\chi_i\}_{i=1}^{N}$ contains the patterns of N classes. Each class $\chi_i = \{\chi_{ij}\}_{j=1}^{N_i}$ with χ_{ij} windows and a set of M feature basis vectors $\{\mho_t\}_{t=1}^{M}$ are estimated by maximizing Fisher's ratio. Fisher's ratio is defined as the ratio of between-class scatter to within-class scatter. The maximization can be formulated as follows,

$$\mho = arg\max\left(\frac{|\mho^T S_b \mho|}{|\mho^T S_w \mho|}\right) \tag{9}$$

where $\mho = [\mho_1, \ldots, \mho_K]$, and S_b and S_w are the between- and within-class scatter matrices, respectively, defined as:

$$S_b = \frac{1}{n}\sum_{i=1}^{N} N_i(\chi_i - \overline{\chi})(\chi_i - \overline{\chi})^T \tag{10}$$

$$S_w = \frac{1}{n}\sum_{i=1}^{N}\sum_{j=1}^{N_i}(\chi_{ij} - \overline{\chi_i})(\chi_{ij} - \overline{\chi_i})^T \tag{11}$$

where $n = \sum_{i=1}^{N} N_i$, is the total number of training windows. The mean of class χ_i is $\overline{\chi_i} = \frac{1}{N_i}\sum_{j=1}^{N_i}\chi_{ij}$. The discriminatory feature vectors can be found corresponding to the vectors of largest eigenvalues. In this experiment, set $\mho$ contains eigenvectors corresponding to k eigenvalues computed from $(S_w)^{-1}S_b$.

3 Experimental Results

The proposed ECG biometric method is tested on MIT-BIH arrhythmia database and QT database of physionet [19]. Both databases include ECG recordings of men and women with the age between 20 and 84 years. The databases have ECG recording of normal and arrhythmia patients. For this experiment, ECG recordings of 48 subjects from MIT-BIH arrhythmia database and 39 subjects of QT database are used. The original sampling rate for MIT-BIH arrhythmia database and QT database is 360 and 250 Hz, respectively. All these records are re-sampled at 200 Hz for this experiment. Each signal is processed with a band-pass filter. Eleven windows of 50 seconds (10,000 samples) and 10 seconds (2000 samples) in length are chosen from processed ECG signal of MIT-BIH arrhythmia database and QT database, respectively. To avoid the sensor and body stabilization effects, the windows are chosen from the middle of each recording. A data set of $528(48 \times 11)\times 10,000$ for MIT-BIH arrhythmia database and of $429(39 \times 11) \times 2000$ for QT database is formed for feature extraction.

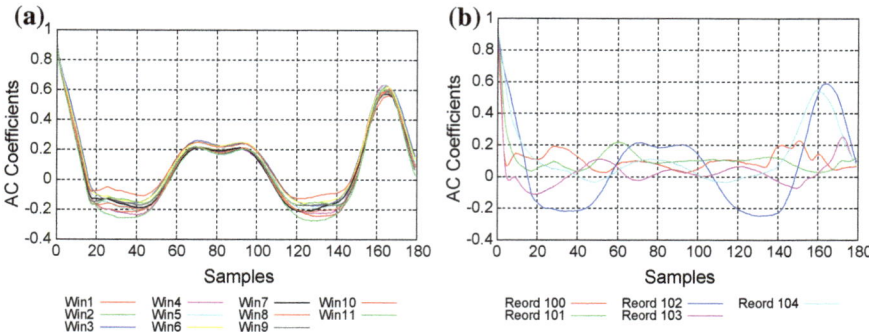

Fig. 2 AC representation of filtered ECG signals: **a** single subject for different windows (eleven) and **b** single window for different subjects (five)

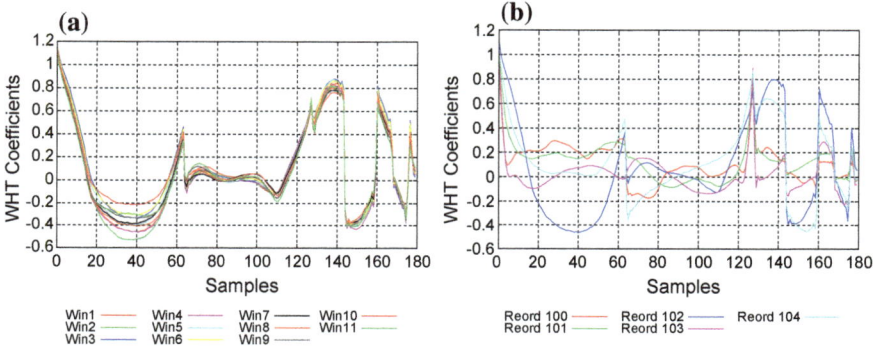

Fig. 3 Plots for Walsh transform of autocorrelated ECG signals: **a** single subject for different windows (eleven) and **b** single window for different subjects (five)

On applying AC to these data sets, the feature vectors of 528×180 and 429×180 are formed for MIT-BIH arrhythmia database and QT database, respectively. The AC time lag of 180 samples is set for this experiment by considering the fact that a normal heart beats 60 to 100 times a minute. The plots of normalized AC for eleven windows of a subject and for five different subjects are shown in Fig. 2a and in Fig. 2b, respectively. These feature vectors are transformed by WHT, in order to minimize the intrasubject variations and to maximize the intersubject variations. The results of WHT for eleven windows of single subject and single window of five different subjects are shown in Fig. 3a and in Fig. 3b, respectively. The LDA is applied for dimensionality reduction of feature vectors to different dimensions such as 2, 5, 7, 10, 13, 15, 20, 22, 25, and 30. The intrasubject variability and intersubject similarity on first three dimensions as achieved by LDA for ten subjects for each database are shown in Fig. 4.

The last window from each record is used as template to form gallery data set. A probe data set is prepared from rest of the windows from each record. The matching

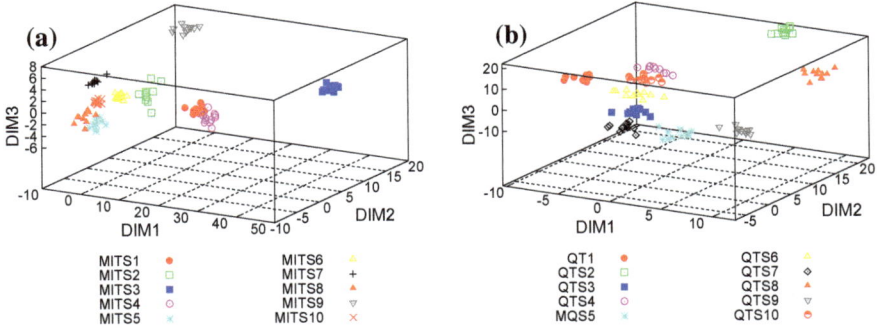

Fig. 4 Intrasubject similarity and intersubject variability represented by first three dimensions as shown by DIM 1, DIM 2, and DIM 3 for ten different subjects of **a** MIT-BIH arrhythmia database and **b** QT database

scores (genuine and imposter) are generated by comparing each projected feature vector from gallery data set to all projected feature vectors in the probe data set. Euclidean distance is used as similarity measure between gallery and probe data sets. The match scores are genuine scores, if they are generated by comparing the attribute sets of probe and gallery data of the same subject; otherwise, the scores are impostor scores. Thus, 48 genuine scores and $2256(48 \times 47)$ impostor scores are generated, for the population of MIT-BIH arrhythmia database. For the population of QT database, the system generates 39 genuine scores and $1482(39 \times 38)$ impostor scores. The performance of the proposed identification system is evaluated using classification accuracy by rank-k. The percentage of probe signals that have the correct class as one of the top k scores is known as the rank-k classification accuracy of the system. Further, the cumulative match characteristic (CMC) curve has been drawn after computing the average rank classification accuracies.

The CMC results at different dimensions for MIT-BIH arrhythmia database are shown in Fig. 5a. The rank-1 classification accuracies of the system at dimensions 10, 13, 15, 20, 27, and 30 are found to be 66, 72, 81, 85, 62, and 60%, respectively. It shows that the rank-1 classification accuracies increase with the increase in dimensions up to twenty (DIM 20) and decreases above DIM 20. The system achieves the better rank-1 classification accuracy of 85% at DIM 20. The CMC curve for DIM 10 shows poor classification performance and reported accuracies of 68% at rank-2, 80% at rank-3, 84% at rank-8, 86% at rank-10, 90% at rank-12, and 100% at rank-36. The classification accuracies at DIM 13 are found to be 80% at rank-3. It increases with the increase in rank and reported as 86% at rank-10, 90% at rank-18, 96% at rank-25, and 100% at rank-29. The classification accuracies are improved at DIM 15 and achieve 100% accuracy at rank-28. The accuracies at other ranks are reported as 85% at rank-2, 91% at rank-9, 97% at rank-25. At DIM 20, the classification accuracies are 91% at rank-2, 95% at rank-10, 97% at rank-19, and 100% at rank-37. The classification accuracies show degradation of performance above DIM 20. At DIM 27, it reports accuracies of 89% at rank-2, 95% at rank-16, 97% at rank-23,

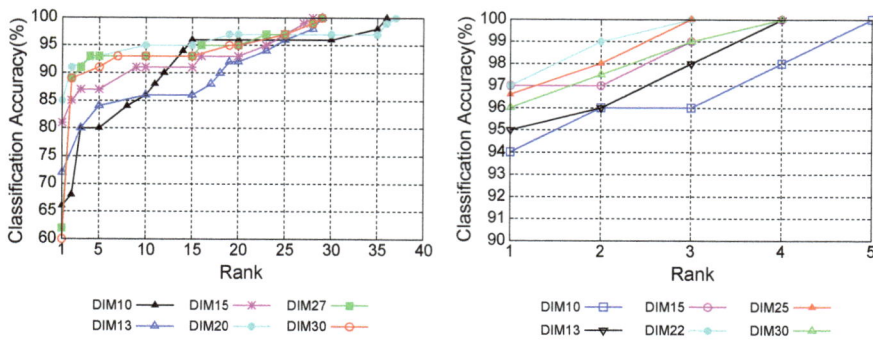

Fig. 5 Cumulative match characteristic curves for rank-based classification accuracies: **a** MIT-BIH arrhythmia database and **b** QT database

and 100% at rank-29. The classification accuracies at DIM 30 are found to be 89% at rank-2, 95% at rank-19, 97% at rank-25, and 100% at rank-29.

The aforementioned system performs better on QT database. The CMC curve is shown in Fig. 5b. The classification accuracies reported at dimension 22 (DIM 22) are found to be 97% at rank-1, 99% at rank-2, and 100% at rank-3. The CMC curve for DIM 10 shows poor performance, and reported accuracies are 94% at rank-1, 96% at rank-2, 98% at rank-4, and 100% at rank-5. At DIM 15, it reports classification accuracies of 97% at rank-1, 99% at rank-3, and 100% at rank-4. The classification accuracies at DIM 30 are found to be 96% at rank-1, 99% at rank-3, and 100% at rank-4.

The average rank classification accuracies on different databases at different dimensions are presented in Table 1. On MIT-BIH arrhythmia database, the average rank classification accuracies are found to be 80, 86, 91, 95, 94, 94, and 91% at dimensions 10, 13, 15, 20, 22, 25, and 30, respectively. The average rank classification accuracies on QT database at dimensions 10, 13, 15, 20, 22, 25, and 30 are reported as 96, 97, 97, 97, 97, 96.6, and 96% respectively. These results reported the highest average rank classification accuracies as 95% at DIM 20 on MIT-BIH arrhythmia database and 97% at DIM 22 on QT database. For both databases, the performance of the system degrades at higher dimensions. For example, above DIM

Table 1 Results of average rank classification accuracies at different dimensions for MIT-BIH arrhythmia database and QT database.

Dimensions	Average rank classification accuracy(%)						
	10	13	15	20	22	25	30
Database							
MIT-BIH arrhythmia database	80	86	91	**95**	94	94	91
QT database	96	97	97	97	**97**	96.6	96

20 on MIT-BIH arrhythmia database and above DIM 22 on QT database, the average rank classification accuracies are linearly decreasing.

These results show that the proposed method reports better identification performance in comparison to the other methods of ECG biometric. For example, the proposed method reports better result than fiducial-based identification method [2]. Although the identification accuracy of 100% was achieved by fiducial point-based methods [6, 7], these methods were tested on only group of 20 subjects. The result of proposed method can also be compared with non-fiducial-based ECG identification methods [4, 11, 12, 15, 16]. Among these, the methods [4, 11, 15, 16] report better performance but they are tested only at 74 healthy subjects, 14 healthy subjects, two sets of 13 subjects each, and 18 subjects only. The proposed method proved to be better in handling the issues like sensitivity to accurate localization of fiducial points of ECG wave and individuality of ECG over larger population.

4 Conclusion

The ECG has emerged as a potential tool for biometric recognition due to its unsusceptibility against spoofing and vitality detection features. The ECG analysis methods based on fiducial points take advantage of different morphological features. Temporal, amplitude, and angle features are significantly different among individuals. These methods rely on accurate localization of fiducial points and their onset and offset. There is no universally acknowledged algorithm to find accurate wave boundaries. This study has analyzed the ECG signal to use as a biometric without detection of its dominant fiducials. The autocorrelation is used to compute the discriminative information available to the ECG signals among population. The autocorrelated signals are transformed into their Walsh coefficients to distinguish the features among them. Further, linear discriminant analysis is used to reduce the dimension of feature vectors to result time- and cost-efficient classification performance. The experimental result has demonstrated that the proposed method of ECG analysis proved to be benchmark for biometric research community as it achieves high identification rate for healthy subjects as well as the subjects having arrhythmia.

Acknowledgements The authors would like to thank the anonymous reviewers and the editor for their feedback and useful suggestions.

References

1. Singh YN, Singh SK (2013) Identifying individuals using eigenbeat features of electrocardiogram. J Eng 2013:8 (Article ID 539284)
2. Singh YN (2014) Individual identification using linear projection of heartbeat features. Appl Comput Intell Soft Comput 2014:14. https://doi.org/10.1155/2014/602813 (Article ID 602813)

3. Pouryayevali S (2015) ECG biometrics: new algorithm and multimodal biometric system, Master of applied science thesis, University of Toronto
4. Gerd W, Manuel S, Dieter K, Ralf-Dieter B, Clemens E (2007) Verification of humans using the electrocardiogram. Elsevier Pattern Recognit Lett 28:1172–1175
5. Jain AK, Ross A, Prabhakar S (2004) An introduction to biometric recognition. IEEE Trans Circuits Syst Video Technol 14(1):4–20
6. Biel L, Pettersson O, Philipson L, Wide P (2001) ECG analysis: a new approach in human identification. IEEE Trans Instrum Meas 50(3):808–812
7. Shen TW, Tompkins WJ, Hu YH (2002) One-lead ECG for identity verification. In: Proceedings of the second joint EMBS/BMES conference, Houston, 23–26 October, pp 62–63
8. Singh YN, Gupta P (2008) ECG to individual identification. In: Proceedings of the 2nd IEEE international conference on biometrics: theory, applications and systems (BTAS 08), pp 1–8
9. Singh YN, Gupta P (2011) Correlation-based classification of heartbeats for individual identi-fication. Soft Comput 15(3):449–460
10. Singh YN (2015) Human recognition using fisher's discriminant analysis of heartbeat interval features and ECG morphology. Neurocomputing Elsevier 167:322–335
11. Plataniotis K, Hatzinakos D, Lee J (2006) ECG biometric recognition without fiducial detection. In: Proceedings of biometrics symposiums (BSYM), Baltimore, Maryland, USA
12. Chan ADC, Hamdy MM, Badre A, Badee V (2008) Wavelet distance measure for person identification using electrocardiograms. IEEE Trans Instrum Meas 57(2):248–253
13. Odinaka I, Lai PH, Kaplan A, O'Sullivan J, Sirevaag E, Kristjansson S, Sheffield A, Rohrbaugh J (2010) ECG biometrics: a robust short-time frequency analysis. In :Proceedings of IEEE international workshop on information forensics and security, pp 1–6
14. Agrafioti F, Hatzinakos D (2008) ECG based recognition using second order statistics. IEEE Computer Society, pp 82–87
15. Wang Y, Agrafioti F, Hatzinakos D, Plataniotis KN (2008) Analysis of human electrocardio-gram for biometric recognition. EURASIP J Adv Signal Process 2008:1–11. https://doi.org/10.1155//148658 (Article ID 148658)
16. Li M, Narayanan S (2010) Robust ECG biometrics by fusing temporal and cepstral information. In: 2010 20th International conference pattern recognition (ICPR), pp 1326–1329
17. Pan J, Tompkins WJ (1985) A real-time QRS detection algorithm. IEEE Trans Biomed Eng 32(3):230–236
18. Beer T (1981) Walsh transforms. Am J Phys 49(5):466–472
19. Physionet, Physiobank archives, Massachusetts Institute of Technology Cambridge. https://www.physionet.org/physiobank/database/#ecg

A Priority Heuristic Policy in Mobile Distributed Real-Time Database System

Prakash Kumar Singh and Udai Shanker

Abstract In fast processing new technology, to provide priority scheduling among different running transactions is a challenging part of research in wireless environment. It incorporates a mechanism to assign priority among transaction to maintain a sequence of execution. Priority heuristics are the backbone of a transaction scheduling approach in real-time systems. It is one of the sophisticated tasks which heavily affect the overall performance of mobile distributed real-time database system (MDRTDBS). Priority heuristics have been developed for centralized and distributed real-time database systems where cohorts or sub-transaction executed in sequential/parallel manner; however, these heuristics may not fit well for the MDRTDBS where sub-transactions are performing parallel execution and face a lot of wireless environment challenges. In this paper, a priority heuristic has proposed which integrates the concept of number of write locks and the variable size of data items. Further, simulation study has been done to evaluate performance of proposed priority heuristics with earliest deadline first and heuristic based on number of locks required.

Keywords Real-time · Transaction · Priority heuristic · Mobile distributed database

1 Introduction

Increasing demand for portable mobile devices, such as laptops, mobile phones, personal digital assistants, and many other wireless devices has become an essential part of our daily life [1–4]. Several real-time applications running on these mobile

P. K. Singh (✉) · U. Shanker
Department of Computer Science and Engineering, MMM University of Technology, Gorakhpur, UP, India
e-mail: pks.cse13@gmail.com

U. Shanker
e-mail: udaigkp@gmail.com

© Springer Nature Singapore Pte Ltd. 2018
M. L. Kolhe et al. (eds.), *Advances in Data and Information Sciences*, Lecture Notes in Networks and Systems 38, https://doi.org/10.1007/978-981-10-8360-0_20

devices such as taxi booking, mobile banking, and stock trading need to complete their transaction within a time constraint. A correct and timeliness method is needed to run these mobile devices properly in intrinsic limitations of wireless environment. In past three decades, research in time-specific distributed real-time database system (DRTDBS) [5–7] received a great attention. To maintain data consistency among transaction, various pessimistic and optimistic concurrency control policies have been developed [8, 9]. However, recent advances in mobile technology introduced a new era of research challenges in the field of mobile distributed real-time database system (MDRTDBS) [8, 9]. MDRTDBS is a collection of mobile and fixed devices (or participants), which are connected through wired and/or wireless channels and share and store the available resources. MDRTDBS performs multiple concurrent transactions which are integrated with a time constraint method.

To perform correct transaction execution and minimize transaction miss rate, researchers have been motivated to develop priority heuristics [4]. Priority heuristics (PH) [4] decide the sequence of transaction's execution. The main concept behind heuristics is to minimize transaction miss rate. So far, many researchers have addressed number of heuristics for centralized and distributed databases. Abbott et al. [1] have addressed transaction scheduling algorithm for centralized real-time database system. They have developed three transaction priority heuristics to achieve data consistency within a time constraint: (i) first come first serve (FCFS), (ii) least slack first (LSF), and (iii) earliest deadline first (EDF). Nowadays, most of the researchers are using EDF with their developed concurrency control policies [8, 9]. Kao et al. addressed the deadline assignment of sub-tasks on various distributed sites [10]. Further, Haritsa et al. [2] have introduced a different approach to decide transaction priority. They have proposed a priority heuristic based on deadline and values that finally resulted in maximizing the actual realized transaction value. In their approach, authors have assigned a specific value to a transaction and expected to have the same value if the transaction finished their commit operation before assigned deadline. Generally, in distributed database, a global transaction (GT) is partitioned into sub-transaction (ST) which runs on different sites.

In last two decades, researchers have developed very few concurrency control (CC) policies in MDRTDBS [8, 11]. Although these CC policies developed for mobile distributed environment, these CCs have also used sequential heuristic methods. These heuristics may not fit well for the MDRTDBS when sub-transactions are performing parallel execution [9, 12]. At first, Shaker et al. [12] have developed a priority heuristic to support parallel sub-transaction execution on different sites for DRTDBS. But it may not fit well in MDRTDBS due to intrinsic limitation of wireless medium. Singh et al. [13] have developed another heuristic using optimistic concurrency control in mobile environment. However, it is not much appropriate for the variable-sized data items distributed among sites [14]. Hence, there is need to develop priority heuristics to fit well for mobile distributed real-time environment where the data item may be of variable size.

Here a new priority heuristic has been proposed to enhance the overall system performance of MDRTDBS which is based on number of write lock required and variable-sized data items. Simulation experiments [13] showed that priority heuristic based on number of write locks outperforms than other existing heuristics over a wide range of workloads. But, they assumed that data items have the fixed (same) size. Most of the systems assume that data items are fixed sized, but in real world, not all data items have the same size. Treating large-sized data items and small-sized data items in the same manner is unfair choice [14]. Ultimately, it is not suitable for variable-sized data items. This motivates us to develop a new heuristic with variable-sized data items.

Instead of past heuristic, where a sub-transaction has been inherited the priority from its parent transaction, our approach assigns priority based on the number of write locks and size of data items required by a sub-transaction at a site. In this paper, performance study has been done by integrated this heuristic with DH2PL [8]. However, the improvement gain as shown by the experimental result is not limited to only DH2PL.

Section 2 briefly describes MDRTDBS transaction model and basic assumptions used in the model. In Sect. 3, proposed heuristic has been detailed. Section 4 presents performance evaluation model parameters and simulation results. Finally, Sect. 5 concludes the paper.

2 MDRTDBS Model

MDRTDBS is a collection of fixed and mobile participants, which can communicate with each other and share resources stored on different sites [8, 9, 15–18]. These participating sites execute multiple concurrent transactions, which should complete before their deadline. The structure of our MDRTDBS model consists of some fixed hosts (FHs), participant mobile hosts (MHs), fixed mobile support stations (MSSs), and different database servers, as given in Fig. 1 [9, 13]. Each MSS is connected through its mobile sites as well as fixed stations. Server site maintains a transaction generator, data manager, concurrency controller, transaction manager, a local database such as main memory, data manager, data broadcasting component [8, 9, 13, 19, 20], ready and wait queue, etc. Transaction generator creates real-time transaction independent to other mobile hosts using Poisson's distribution using given arrival time. Mobile site consists of a mobile transaction generator, mobile concurrency controller, mobile transaction manager, CPU, ready and wait queues for mobile transactions, a wireless communication interface, a local database, a sink, data manager, data receiver, data manager, and disconnection processor. In this model, the transaction is initialized at mobile clients within its cell or from others cell. The fixed server (MTSO) is responsible for active transaction maintenance, channel allocation, handoff procedure, message routing and allocation of cohort on different base stations (Fig. 1).

Mobile data manager manages the locally committed data items of read-only transaction (ROT) [16, 18, 20, 21]. Data receiver receives the committed data broadcasted by server in last broadcast cycle. In this paper, mobile transaction can be initiated at mobile host. Transaction from mobile host is assumed to be flat transactions which allow only firm deadline. If the transaction misses their dead-line, it must be aborted. Each transaction is assigned an initial deadline. In real-time scenario, it is assumed that the transactions have their predefined set of read and write data items. Each base station and server maintains a lock table, and the data conflict is detected by a scheduling mechanism. In our MDRTDBS model, the most important component is priority assignment which assigns the priority to different mobile transactions. Priority assignment policy specifies the order of execution of mobile transactions. Based on transaction priority, the mobile sub-transactions known as cohorts are queued in ready queue and it must pass through concurrency control mechanism to obtain a lock on the data item. The MDRTDBS model uses two-phase commit (2PC) to perform its commit procedure. In the given MDRTDBS model,

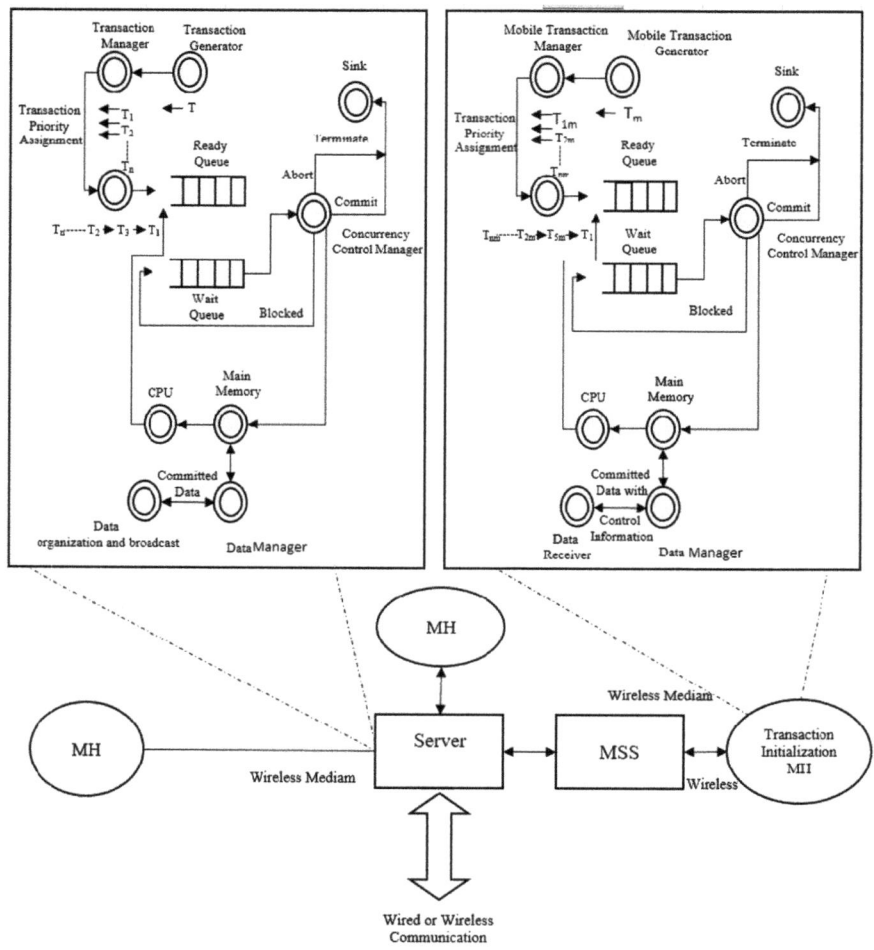

Fig. 1 Mobile distributed real-time database model

sink component gathers the statistics for all committed or aborted transactions. After commitment procedure, transactions release their read or write locks.

3   Proposed Priority Heuristic

The new heuristic is based on number of write locks required by the cohorts with addition to total size of data items of transaction on which locks are required. Initial priority (Init_P) of each cohort T_i is computed using following formulae:

$$\text{Init\_P} = 1/N_w + 1/S \quad \text{where,} \quad N_w \geq 1 \tag{1}$$

where it is assumed that N_w is the number of write locks required by the cohort and S is total size of data items of transaction on which locks are required. In mobile environment, the percentage of read-only transaction is much more than update transactions [9, 16, 18]. The write locks of update transaction have a greater chance of confliction than read locks, and it takes much processing and communication time than read locks [8]. So, it is not fair to consider only fixed-sized data items. In this paper, the size of data items is assumed to be variable. In deciding initial priority of cohort transactions, the concepts of total size of data items have been added with number of write locks. An intermediate priority assignment policy (I_Pr) is added with the proposed heuristic to minimize the transaction miss rate.

I_Pr is used with the proposed heuristic to minimize the restart rate. In this policy, an intermediate priority is assigned to newly arrived lock requesting cohort. Basically, this policy prevents the unnecessary abort of lock holding low priority cohort T_{lh}. This scheme is primarily based on the total remaining execution time Remain (T_{lh}) required for lock holding low priority cohort T_{lh} and the available slack time ST (T_{lr}) of the newly arrived higher priority cohort T_{lr}. Even if the requesting arrived cohort T_{lr} is higher priority than T_{lh}, the low abort of T_{lh} is possible only when the slack time of T_{lr} is lower than the total remaining execution time of T_{lh}. In this procedure, this policy does not affect the initial priorities assigned to the two cohorts. The total remaining execution time of the lock holding cohort (T_i) is given as:

$$\text{Remain } (T_i) = R_i - \text{Elapse } (T_i)$$

where assume that Remain (T_i), Elapse (T_i), R_i, ST (T_i) are the remaining time left of transaction execution, elapsed time of transaction execution, remaining time needed of transaction, and slack time of T_i, respectively. In case of Remain, (T_{lh}) is less than ST (T_{lr}) and then T_{lr} waits for the lock which holds by executing T_{lh}; otherwise, it forces T_{lh} for performing the abort procedure. The following is the algorithm developed for the above-proposed heuristic with I_Pr. Assume that Pr (T_i) is the priority of cohort T_i.

Algorithm:
For each: Arrived T_{lr}
{

    Assign Init_P to T_{lr};

}
For each: data items
{

    If(! Lock_conflict)
    {

        Allocate data item to T_{lr};

    }
    Else if (Pr (T_{lr}) > Pr (T_{lh}) and T_{lh} is not committing)
    {

        For each: data items \\T_{lh} *a Local or global Transaction*
        {

            If(ST(T_{lr})< Remain (T_{lh}))
            {

                Abort T_{lh};
                Allocate data item to T_{lr};

            }
            Else
            {

                Insert T_{lr} in wait queue;

            }
        }

    }
    Else if (Pr(T_{lr})> Pr(T_{lh}) and T_{lh} is committing)
    {

        Wait T_{lr} until T_{lh} unlock the locks;
        Pr (T_{lh}) = Pr (T_{lr}) + Threshold value;

    }
    Else if (Pr (T_{lr}) <= Pr (T_{lh}))
    {

        Wait T_{lr} until T_{lh} releases the lock;

    }

}

4 Performance Evaluation

4.1 Performance Parameters and Measures

The MDRTDBS having N_{sites} different sites is simulated based on [8, 9, 12, 14]. Table 1 consists of the default values of related parameters. The parameter has been taken based on earlier simulator developed by Lee et al. [4, 14], Lam et al. [8], Lei et al. [9], and Shanker et al. [12]. Simulation has been written using C language. The transaction is assumed to be initiated at mobile site. Server distributes the cohorts among different sites. After completion of the transaction on different sites, the next transaction will generate using think time given in Table 1. The concurrency control policy DH2PL is used to evaluate the impact of EDF, heuristic of [12], and proposed heuristic with I_Pr policy. In this paper, it has been assumed that m is number of cohorts of transaction T, N is number of operations, T_l is time required to lock or unlock a data item, T_{pr} is the time to process a read or write data item, T_{comm} is communication delay between server and mobile node, N_{comm} is number of messages communicated between server and mobile host, T_{TPT} is total processing time of local or global transaction, and SF is the slack factor. Then, estimated deadline of a transaction Y is calculated as:

$$DL(Y) = AT(Y) + (T_{TPT} \times N + T_{comm} \times N_{comm}) \times (1 + SF)$$

For local transaction, value of $= 0$ and T_{TPT} is calculated as,

$$T_{TPT} = 2T_l + T_{pr}$$

However, for global transaction and for m cohorts, the value of T_{TPT} is computed as:

$$T_{TPT} = 2T_l + T_{pr} \times m.$$

Table 1 maintains the baseline setting for the simulation work. The parameters are chosen to create a scenario with high utilization and variable-sized data items. In our simulation work, decreasing distribution (DEC) [14, 22, 23] is used for data access pattern, where default value of θ is taken as 0.8. A mobile transaction is processed until it is committed. Once the transaction deadline is found missing, the transaction will be aborted.

Miss rate and restart rate are two performance metrics of our simulation. Following formulae is used to determine the miss rate and restart rate:

$$Miss\ Rate = \frac{number\ of\ deadline\ missing\ transaction}{total\ number\ of\ transactions\ submitted\ for\ processing}$$

Table 1 Baseline setting

Parameter	Meaning	Default value
DB_SIZE	Database size	500
NL_DB_SIZE	No of local database size	5
d_size Min	Minimum size of data item	5 Kbytes
d_size Max	Maximum size of data item	1000 Kbytes
B	Bandwidth	128 K bytes per second
DT	Disk time	20 ms
PHR	Hit ratio	75%
CPU_P	Time to process an operation	10 ms
CPU_L	Time to lock a data item	1 ms
CPU_{RL}	Time to unlock a lock	1 ms
CPU_{UD}	Time to write a data item	4 ms
L_UI	Location update interval	0.2 s
θ	Zipf distribution parameter	0.8, range is between 0 to 1
P_WL	Proportion of write lock	0.6
S_R	Slack range	2–6
N_CELL	Number of mobile cells	5
N_MC	Number of mobile clients	10
Pdiss	Probability of disconnection	0.5%
Tdis	Duration of disconnection	1–50 s
P_ROT	Proportion of ROT	70%
P_H	Handoff probability	2%
T_T	Think time	1–7 s

$$\text{Restart Rate} = \frac{\text{Total number of local (or global) transaction estart}}{\text{total number of local (or global) transaction submitted}}$$

4.2 Simulation Result

In this simulation work, a comparison has been performed among our heuristic (HP_NWL_SD), basic EDF-based policy (HP_EDF), and a heuristic proposed by Shanker et al. (HP_NL) [12]. Increase in think time ultimately decreases the transaction workload. Hence, in Fig. 2, the graph shows the transaction miss rate corresponding to the think time and it seems that an increase in think time decreases the miss rate. Similarly, in Fig. 3, the graph shows the restart rate decreases with respect to increase of think time. In our simulation work, study the performance of the proposed heuristic with high data contention, a Zipf distribution of data items (default value = 0.8) in global as well as local transactions. The proportion of write

Fig. 2 Miss rate versus think time

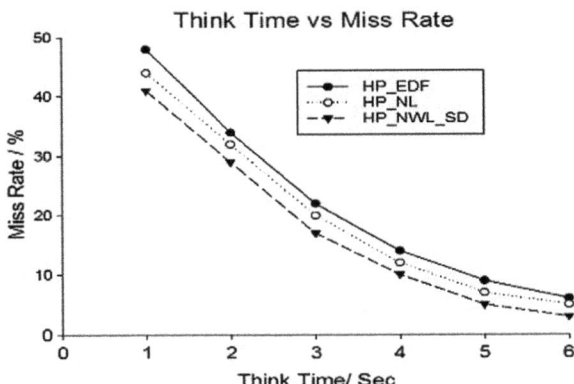

Fig. 3 Restart rate versus think time

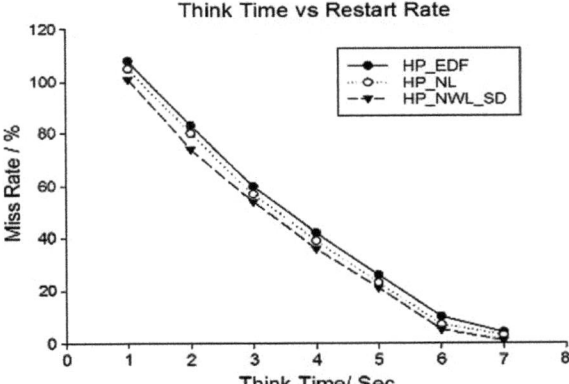

lock has been taken as 0.6 and communication disconnection as 1–50 s. Locks are granted in a controlled manner.

The miss rate and restart rate have been calculated using the formulae mentioned in last Sect. 4. In Fig. 2, its look clearly that the performance of the proposed approach is consistently better than the other two policies (as shown in figure). Similarly, in Fig. 3, using restart rate metric, it has been found that the proposed approach HP_NWL_SD performs better than PH_NL and PH_EDF.

5 Conclusion

Priority heuristic policy for mobile distributed real-time database systems (MDRT-DBS) is a backbone for MDRTDBS. This paper proposes a priority heuristic for MDRTDBS which includes two main factors, i.e., number of write locks and total size of data items. Simulation results show that these factors are critical to performance of CC protocol and suitable in mobile environment. The proposed heuristic has been introduced using intermediate priority assignment policy. Many heuristics have been addressed, but they may not be fit in mobile environment. The proposed heuristic is compared with HP_EDF and HP_NL, and series of simulation work indicates that the proposed heuristic outperforms well than other priority assignment policies. Further to minimize transaction miss rate, other heuristic approaches based on several issues of wireless medium could be possible to achieve transaction consistency in mobile environment.

References

1. Abbott RK, Molina HG (1992) Scheduling real time transactions: A performance evaluation. ACM Trans Database Syst 17(3):513–560
2. Haritsa JR, Carey MJ, Livny M (1992) Data access scheduling in firm real-time database systems. J Real-Time Syst 4(3):203–242
3. Lam KY, Lee VCS, Hung SL, Kao BCM (1997) Priority assignment in distributed real-time databases using optimistic concurrency control. IEE Proc-Comput Digital Tech 144(5):324–330
4. Lee VCS, Lam KY, Kao BCM, Lam KW, Hung SL (1996) Priority assignment for subtransaction in distributed real-time databases. In: First international workshop on real-time database systems (1996)
5. Lam KY (1994) Concurrency control in distributed real-time database systems. Ph.D. thesis, Department of Computer Science, City University of Hong Kong, Hong Kong
6. Shanker U, Misra M, Sarje AK (2006) SWIFT: a new real time commit protocol. Distrib Parallel Databases 20(1):29–56
7. Shanker U, Misra M, Sarje AK (2008) Distributed real time database systems: background and literature review. Int J Distrib Parallel Databases 23(2):127–149
8. Lam KY, Ku TW, Tsang WH, Law GCK (2000) Concurrency control in mobile distributed real-time database. J Inf Syst 25(4), 261–286
9. Lei X, Zhao Y, Chen S, Yuan X (2009) Concurrency control in mobile distributed real-time database systems. J Parallel Distrib Comput 69:866–876
10. Kao B, Molina HG (1993) Deadline assignment in a distributed soft real-time system. In: Proceedings 13th international conference on distributed computing systems, pp 428–437
11. Xiangdong L, Yuelong Z, Songqiao C, Xiaoli Y (2010) A multiversion optimistic concurrency control protocol in mobile broadcast environments. Int J Comput Appl 32(3):261–266
12. Shanker U, Misra M, Sarje AK (2005) Priority assignment heuristic to cohorts executing in parallel. In: Proceedings of the 9th WSEAS international conference on computers, World Scientific and Engineering Academy and Society (WSEAS), pp 1–6
13. Singh PK, Shanker U (2017) Priority heuristic in mobile distributed real time database using optimistic concurrency control. ADCOM 2017
14. Lee VCS, Wu X, Ng JKY (2006) Scheduling real-time requests in on-demand data broadcast environment. J Real-Time Syst 34(2):83–99
15. Pitoura E, Chrysanthis PK (1999) Scalable processing of readonly transactions in broadcast push. In: Proceedings of the 19th IEEE international conference on distributed computing system, pp 432–439
16. Lee VCS, Lam KW, Son SH (2000) Real-time transaction processing with partial validation at mobile clients. In: Proceedings of seventh international conference on real-time computing systems and applications. IEEE, pp 473–477
17. Lee VCS, Lam KW, Son SH, Chan EYM (2002) On transaction processing with partial validation and timestamp ordering in mobile broadcast environments. J IEEE Trans Comput 51(10):1196–1211

18. Lee VCS, Lam KW, Kuo TW (2004) Efficient validation of mobile transactions in wireless environments. J Syst Softw 69(1):183–193
19. Herman G, Lee KC, Weinrib A (1987) The datacycle architecture for very high throughput database systems. Proc ACM SIGMOD Record 16(3):97–103
20. Shanmugasundaram J, Nithrakashyap A, Sivasankaran R, Ramamritham K (1999) Efficient concurrency control for broadcast environments. ACM SIGMOD Record 28(2):85–96
21. Park S, Jung S (2009) An energy-efficient mobile transaction processing method using random back-off in wireless broadcast environments. J Syst Softw 82(12):2012–2022
22. Hameed S, Vaidya NH (1999) Efficient algorithms for scheduling data broadcast. ACM/Baltzer J Wireless Network 5(3):183–193
23. Zipf PGK (1949) Human behavior and the principle of least effort. Addison-Wesley, Massachusetts
24. Qin B, Liu Y (2003) High performance distributed real time commit protocol. J Syst Softw, pp 1–8 (Elsevier Science Inc)

A Proposal for Optimization of Horizontal Scaling in Big Data Environment

Chandrima Roy, Manjusha Pandey and Siddharth Swarup Rautaray

Abstract The data which is beyond the storage space of the server and beyond to the processing power is called Big Data. It is not manageable by traditional RDBMS or conventional statistical tools. Big data increases the storage capacities as well as the processing power. Horizontal scaling or sharding is needed to divide the data set and distributes the data over multiple servers. Redundancy and fault tolerance are achieved by horizontal scaling. Optimization of horizontal scaling is an important aspect of Big Data technology. Instead of using vertical scaling that means upgrading to fancier computers when the current system becomes inadequate, we have to add more node (computers) to a cluster. It increases the parallelism, rather than the performance of any one node. This paper presents the fundamentals of big data analytics but directing toward an analysis of various optimization techniques used in the big data environment.

Keywords Optimization · Horizontal scaling · RDBMS · Hadoop · Big data tools

1 Introduction

Big Data involves the data produced by different devices and applications. Big data is generated from everything around us. Big data is a concept which actually explains about gathering, organizing, analyzing, and getting the information out the data and also the ability to store such amount of data [1]. The major benefit of using big data is *that data collected from different sources is in smaller chunks so the program can process them in a parallel manner with a great speed.* Optimization is needed

C. Roy (✉) · M. Pandey · S. S. Rautaray
School of Computer Engineering, KIIT University, Bhubaneswar, India
e-mail: chandrima.roy.1914@gmail.com

M. Pandey
e-mail: manjushafcs@kiit.ac.in

S. S. Rautaray
e-mail: siddharthfcs@kiit.ac.in

© Springer Nature Singapore Pte Ltd. 2018
M. L. Kolhe et al. (eds.), *Advances in Data and Information Sciences*, Lecture Notes in Networks and Systems 38, https://doi.org/10.1007/978-981-10-8360-0_21

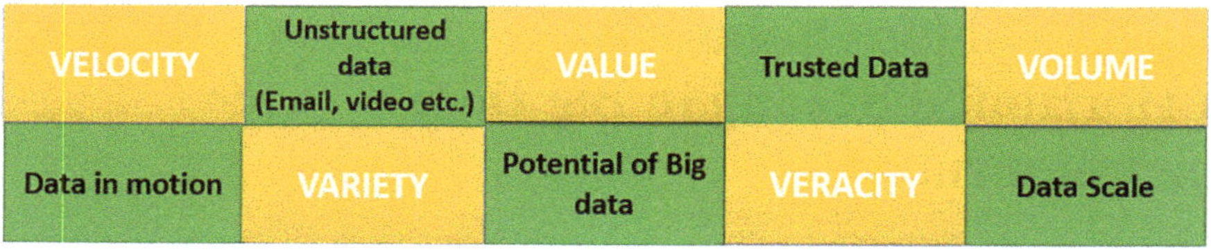

VELOCITY	Unstructured data (Email, video etc.)	VALUE	Trusted Data	VOLUME
Data in motion	VARIETY	Potential of Big data	VERACITY	Data Scale

Fig. 1 Big data characteristics

to handle different workloads and integrates them into a single infrastructure. The ground rule of optimization is to never assume anything, always verify using actual data set (Fig. 1).

Volume is the amount of data generated every day. The sources may be internal or external. Velocity is how fast data is processed, batch and streaming data. Variety deals with a wide range of data types and sources of data. This is generally studied under three categories: structured data, semi-structured data, and unstructured data. Veracity refers to the uncertainty of data, i.e., whether the obtained data is correct or consistent. Out of the huge amount of data that is generated in almost every process, only the data that is correct and consistent can be used for future analysis. Value is the identification of data which is valuable than transformed and analyzed.

Optimization of big data prepares the logical schema from the data view schema [2]. Optimization is an important feature in database management and in the data warehouse. Big data optimizations are done to fetch data from a data sources so that the data can be used for various purposes. Optimization of horizontal scaling is needed to minimize the execution time and to maximize the performance of the system.

2 Background

As increasing volume of data with large-scale and complexity is being erected progressively, identifying structured data from unstructured big data in a scalable manner has become a big challenge. As the data is noisy, sparse, and heterogeneous, the big data challenge is also aggravated. Table 1 presents some papers about optimization of horizontal scaling in Big Data environment which illustrates the benefits of using an optimization framework big data analytics.

Esma Yildirim et al. proposed a paper about pipelining, parallelism, and concurrency which are used for application-level optimization. Factors that affect the data transfer throughput are network characteristics, end-system characteristics, dataset characteristics. Some mechanisms are proposed to overcome data transfer bottlenecks. Their optimal values depend on available bandwidth, RTT, CPU, and disk speed as well as the transfer characteristics. Effects of these parameters are described in this paper, which has an effect on the throughput of large dataset transfers, and it

Table 1 Study and getting idea from different papers

Proposed mechanism	Description	Details
Hadoop performance optimization	Hadoop platform provides an improved programming model, which is used to create and run distributed systems quickly and efficiently. This paper describes the Hadoop framework, optimization work on MapReduce, and the performance of HDFS. Advantages and disadvantages of these technologies are also described by the author	Feng et al. [1]
Self-optimization of handover technique using big data analytics	Big data evaluate the operation of a cellular network. Automatic handover self-optimization is presented in this paper using big data. In each cell of the network, self-optimization of NCL is performed to achieve handover self-optimization	Lee et al. [3]
Optimization business process with big data analytics	One of the contemporary problems in a business network is the vast amounts of data generated from various sources. The supply chains will need administration for support of privacy aspects of cooperating business units existing in big data ecosystems	Robak et al. [4]
Real-time anomaly evolution monitoring using big data analytics	Micro-blogging has been very popular in the big data era due to its real-time scattering of information. It is important to monitor what is trending on the social media. In this paper, author proposed a real-time emerging anomaly monitoring system (RING)	Yu et al. [5]

also provides several techniques. Maumita Bhattacharya et al. proposed a paper about evolutionary optimization and stochastic search techniques. Evolutionary algorithms (EA) explore search space including deterministic methods. The proposed model introduces diversity by using informed genetic operators. The paper also introduces an algorithm to deal with the high-dimensionality problem.

3 Proposed Framework

Scalability refers to the system's capacity to expand the computer resources to handle the exponential growth of work by increasing the total output when resources are added [2]. If the database is not scalable, then the processes can slow down or even fail. Scalability enables the database to grow to a larger size to support more transactions. **Scale-out or horizontal scaling** is done to increase the capacity by connecting multiple software or hardware entities in such a manner that they function as a single logical unit, rather than making a node more powerful, and many nodes are added to the cluster. The term scaling out is also used for horizontal scaling [3]. To take advantage of horizontal scaling, we need specialized program that is designed to take advantage of distributed system that can pass the processing to multiple computers and that can store the data on multiple computers. **Horizontal scaling is** cheaper and fault-tolerant process that can be applied to smaller systems [4].

Figure 2 describes the Hadoop cluster where optimization is done by including the Zookeeper to control the overall distribution. Previously, the number of Data Node is limited so optimization is also not needed, but periodically with the increase of the huge amount of data, the number of Data Node also increases to store all the data, so optimization is needed. Name node controls the HDFS [6]. It does not hold any actual file data, rather it holds all file system metadata for the cluster. Such as name of the file and its path, what blocks makeup a file and where those blocks are located in the cluster. It holds the health of each and every Data Node [7]. Backup node or secondary name node connects to the name node (occasionally) and takes a copy metadata store into NameNode's memory [8]. If the NameNode dies, the files preserve by secondary NameNode used to recover the NameNode.

Data Node sends heartbeats to the NameNode every 3 s to inform the NameNode that it is still alive. Every 10th heart beat is a block report, Data Node tells the NameNode about all the blocks contains within it. Node Manager takes care of the computational nodes in the Hadoop cluster. It determines the health of the executing nodes. It keeps a track of the resources available and sends the track report to resource manager [9]. Journal Node is used to keep the state synchronized between the NameNode and Backup Node or standby NameNode. When any modification is performed by the NameNode, it informs the journal node about the modification. The secondary node is able to read the update from the Journal node (Fig. 3).

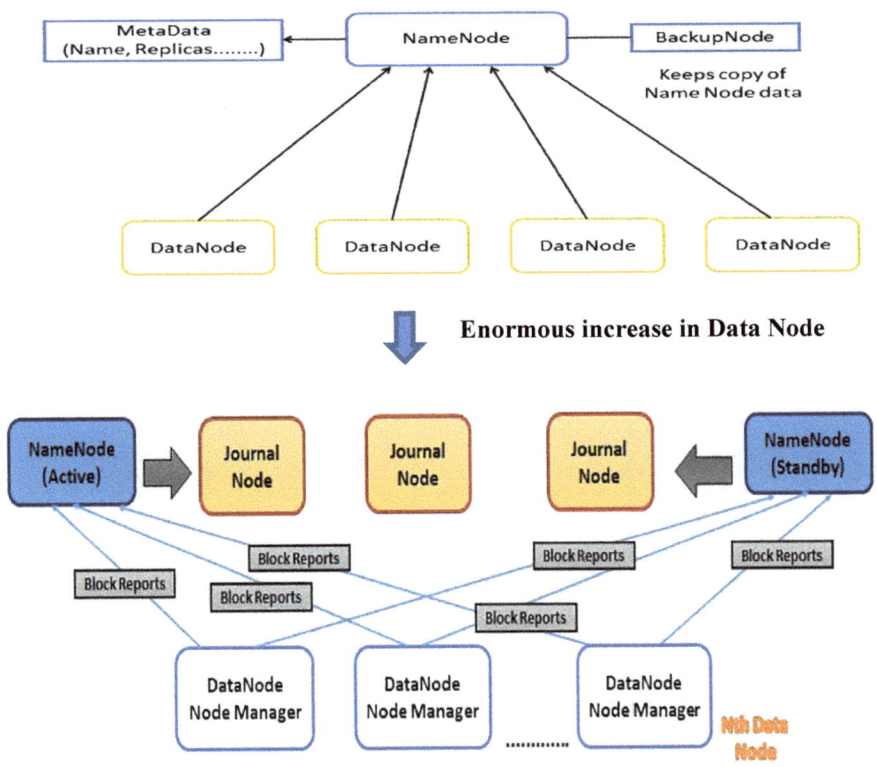

Enormous increase in Data Node

Fig. 2 Optimization of Data Node

Fig. 3 Flowchart of the proposed framework

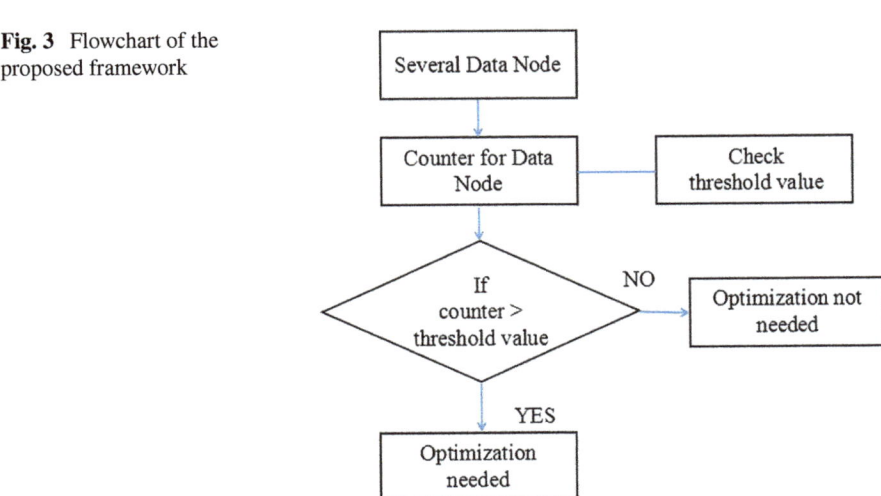

3.1 Proposed Algorithm for Horizontal Scaling Optimization

```
Input:  -  Number of Data Nodes -> N(D_N)
           Task Assign to each data node -> (T_DN)
           Total task assign -> N(T_DN)
Step 1:-  J(N_N) -> A Job assign to Name Node
Step 2:-  f_T -> { ∀ T_x ∈ J(N_N)
           Let T_x= {T_1, T_2, T_3, ....T_N} }
           Where f_T is the function which breaks the Job
           J(N_n) to various task T_x that is a set of T_1,T_2
           T_3,....T_N.
           Each node represented by a count value.
Step 3:-  f_TDN -> Assign { T_1 -> T_DN1
                            T_2 -> T_DN2
                            T_3 -> T_DN3
                            .
                            .
                            T_N -> T_DNN}
           Where f_TDN is the function which assign each
           task to each data node.
Step 4:-  Conditional statement
           C_1= {If T_DNN > Threshold}
           Here we are checking whether the T_DNN value is
           greater than the threshold value or not.
Step 5:-  If C_1 == True
           f_o -> optimize(f_TDN)
           else
           Follow f_TDN
Output:   Case 1 ==> Optimization needed.
           T(N_5)-> N_N
           Leftover task assign to Name Node rather
           than using another Data Node for execution.
           Case 2 ==> Optimization not needed.
                      Task assign to each data node.
```

Optimization of horizontal scaling can be done by optimizing the use of Data Node. The above algorithm represents how task management is done by Name node (N_N).

$N(D_N)$ is the number of Data Nodes present in the execution. NameNode assigns task to each Data Node which is represented by (T_{DN}). So the total assigned task is $N(T_{DN})$. A Job $J(N_N)$ is assigned to NameNode. A function f_T breaks the Job $J(N_N)$ into various tasks T_x that is a set of $T_1, T_2, T_3,...T_N$. f_{TDN} is the function which assigns each task to each Data Node. A counter variable is present to count the number of Data Node used in a certain execution [5] which is represented by T_{DNN}. A threshold value is set from previously to check the counter value with it. The result of the checking decides whether optimization is needed or not. That means if the counter value is greater than the threshold value, then only optimization is needed.

Table 2 Big data components

S. No.	Big data components	Description
1	Hadoop	Hadoop is an open-source platform that provides analytical technologies and computational power required to work with such volumes of data. Hadoop platform provides an improved programming model, which is used to create and run distributed systems quickly and efficiently
2	MapReduce	Hadoop accomplishes its operations with the help of the MapReduce model, which comprises two functions, namely a mapper and a reducer. The mapper function is responsible for mapping the computational subtasks to different nodes and also responsible for task distribution, load balancing, and managing the failure recovery. The reducer function takes the responsibility of reducing the responses from the compute nodes to a single result. The reducer component has the responsibility to aggregate all the elements together after the completion of the distributed computation
3	HDFS	HDFS is an effective, scalable, fault tolerant, and distributed approach for storing and managing huge volumes of data. The data collected in a Hadoop cluster is first broken down into smaller chunks called blocks and distributed across multiple nodes. These smaller subsets of data are then operated upon by mapper and reducer functions. HDFS works on the write once read many times approach

4 Hadoop

The core Big Data components can be classified as—Hadoop Ecosystem, MapReduce, HDFS, Cloudera, MongoDB, NOSQL databases, etc. The classification of Big Data tools is based on the fact that the abovementioned technologies are employed to address big data requirements. Following are the various Big data tools discussed in this paper (Table 2).

5 Conclusion

This paper discusses about how we can optimize horizontal scaling in a Big Data environment. It has presented some idea about various tools like HADOOP, MAPRE-DUCE, and HDFS which are required to process the big data. A fast transfer of huge amount of data to a slow computing can be the reason of data overflow, whereas slow transfer of data to relatively fast computing nodes can cause data underflow. Computing nodes will remain idle over the time in case of data underflow. So according to the overflow and underflow condition, optimization is needed. It proposes an optimization technique to reduce the workload from Data Nodes. It can be done in various ways. Suppose we have data which can be managed by four Data Node but still some data is left, for which we can use another Data Node. But that last Data Node is not utilized fully. Instead of using the fifth Data Node, the smallest amount of data could have been analyzed by the name node or some auxiliary node. Assuming we have several computers (Data Node), the current computer specifications to ensure the ability to operate are: Processor—Intel(R) Core(TM) i3 CPU, RAM—4 GB, System type—32-bit operating system, and Windows 7 Professional. We have to use this machine to complete some specific task, which can be scaled horizontally. This paper proposes the mechanism of optimization of horizontal scaling. We can use the concept of auxiliary node in order to optimize horizontal scaling. In recent years, there has been a boom in Big Data because of the growth of different kind of data coming from different internal and external sources. We now have huge amounts of data, and it is up to different organizations to exploit the data in order to extract useful information [3]. The presented paper represents the various applications of big data, enhancement of horizontal scaling, zookeeper to provide distributed synchronization and optimization techniques used in the big data environment. The paper concludes that the future research work is focused on those areas.

References

1. Feng D, Zhu L, Zhang L (2016) Review of hadoop performance optimization. In: 2016 2nd IEEE international conference on computer and communications (ICCC). IEEE
2. Westwood JA, Cazier JA (2016) Work-Life optimization: using big data and analytics to facilitate work-life balance. In: 2016 49th Hawaii international conference on system sciences (HICSS). IEEE
3. Lee CL, Su WS, Tang KA (2014) Design of handover self-optimization using big data analytics. In: 2014 16th Asia-Pacific network operations and management symposium (APNOMS). IEEE
4. Robak S, Franczyk B, Robak M (2012) "Applying linked data concepts" in BPM. In: FedCIS 2012, IT4L. IEEE Conference Publications, pp 1105–1110
5. Yu W, Li J, Bhuiyan MZA, Zhang R, Huai J (2017) Ring: real-time emerging anomaly monitoring system over text streams. IEEE Trans Big Data PP(99):1
6. McCreadie R, Macdonald C, Ounis I, Osborne M, Petrovic S (2013) Scalable distributed event detection for twitter. In: 2013 international conference on IEEE big data
7. Yang HC, Dasdan A, Hsiao RL (2007) Map-reduce-merge: simplified relational data processing on large clusters. Sigmod 1029–1040
8. Banerjee A, Bandyopadhyay T, Acharya P (2013) Data analytics: Hped up aspirations or true potential? Vikalpa 38(4):1–11
9. Boyd D, Crawford K (2012) Critical questions for big data. Inform Commun Soc 15(5):662–679

A Literature Review on Hadoop Ecosystem and Various Techniques of Big Data Optimization

Vikash Kumar Singh, Manish Taram, Vinni Agrawal
and Bhartee Singh Baghel

Abstract We are living in twenty-first century, and this century means for its faster work, accurate analysis, highly processed data, and speed. This is the epoch of "Big data." Big data is a term that describes huge mass of structured and unstructured data that is unable to be processed by traditional data processing systems. Big data stands for storage of large amount of data to extract the valuable content with its characteristics 5-Vs, i.e., Volume, Variety, Velocity, Veracity, and Value. But before the arrival of Hadoop, procuring and depository of data was an issue. Hadoop takes its first step in the Data Science Market in 2005. It was created by Doug Cutting and Mike Cafarella. Hadoop is a software framework that allows users to depot data and run their applications on Hadoop clusters. Its best part is its open-source framework.

Keywords Hadoop · Big data · MapReduce · Pig · Hive · Sqoop

1 Introduction

Big data comes as a boon to the industries flooding with enormous amount of data. This huge quantity gives an idea about how much software related Big data a large enterprise sits over [1]. The traditional data analytics may not be able to handle such large quantities of data [2]. There is a strong need to have a methodology specifically for Big data projects [3]. Big data processing presents new opportunities due to its analytic powers [4] (Table 1).

V. K. Singh · M. Taram (✉) · V. Agrawal · B. S. Baghel
IGNTU, Amarkantak, India
e-mail: manishtaram86@gmail.com

V. K. Singh
e-mail: drvksingh76@gmail.com

V. Agrawal
e-mail: vini8425@gmail.com

B. S. Baghel
e-mail: bharti1926singh@gmail.com

© Springer Nature Singapore Pte Ltd. 2018
M. L. Kolhe et al. (eds.), *Advances in Data and Information Sciences*, Lecture Notes
in Networks and Systems 38, https://doi.org/10.1007/978-981-10-8360-0_22

Table 1 Areas getting benefited by use of Big data

Sectors	Areas using Big data
Retail sector	• Detecting their store locations • Recording customer's opinions on pricing
Business sector	• Product research • Quality assessment
Financial services	• Risk analysis • Fraud and mischief prevention
Government	• Analyzing economic status • Marketing strategies
Healthcare sector	• Bioinformatics • Pharmaceutical research
Advertising companies	• Demand signaling • Tracking advertising
Media and telecommunications sectors	• Computing customer reviews • Streaming optimization

2 Literature Review on Hadoop Operating System

The storage portion of the Hadoop framework is provided by a distributed file system solution such as HDFS [5]. HDFS is the main component of Hadoop. It runs on commodity hardware and provides easier access and storage of structured, semi-structured, and unstructured data on its clusters. Compression capabilities in Hadoop are limited because of the HDFS block structure [6]. HDFS is highly scalable and advantageous in its portability [7]. HDFS divides data into multiple blocks along with their replications which makes its fault tolerant. In addition to exploiting concurrency of large numbers of nodes, HDFS minimizes the impact of failures by replicating data sets to a configurable number of nodes [8]. HDFS works in master–slave architecture of Hadoop and provides parallel processing of applications. Once a file is written in HDFS, it can be read as many times as any authenticate user wants too; hence, HDFS is also secured. HDFS splits a file into a small size of 64 MB. Hadoop cluster is a type of computational cluster being used for storing and processing masses of unstructured data in the environment of distributed computing.

3 Master–Slave Architecture of Hadoop

Table 2 shows name node which is used by master services of Hadoop for storing file's metadata, for monitoring the coordination access of data stored, and keeping a record of system information. Secondary name node achieves data from name node and forwards it further for analyzation after keeping its replication copy for future circumstances.

Table 2 Components of HDFS and their descriptions

HDFS components	Job	Working level
Name node	• Executes operations of file systems (closing file, opening file, etc.) • Regulates file's access to the clients	Master level
Secondary name node	• Contacts name node in periodic manner for assigned task • In case of name node failure, secondary name mode takes its place and updates it using fsimage file	Master level
Job tracker	• Manages processing of data files with the help of MapReduce • Allot processing times and criteria to specified jobs	Master level
Data node	• Stores chunks of data and retrieve them in respected time • Perform read and write requests as per the instructions of name node	Slave level
Task tracker	• Runs MapReduce jobs on files provided by name node • Informs the current status of running task to name node	Slave level

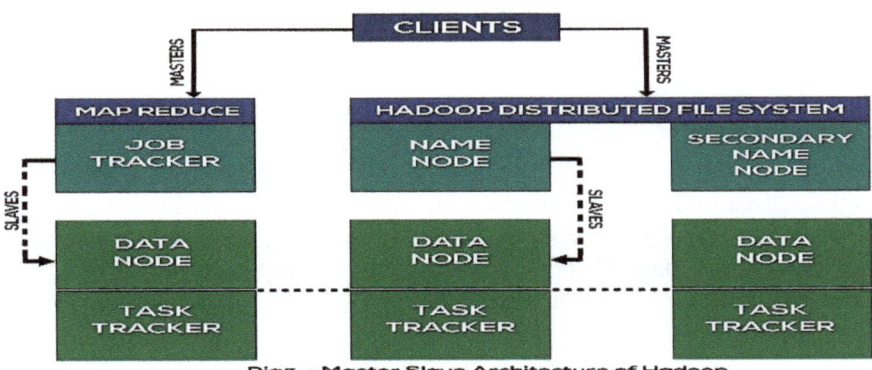

Diag. – Master Slave Architecture of Hadoop

Fig. 1 Master–slave architecture of Hadoop (Reference https://data-flair.training/blogs/hadoop-ecosystem-components/)

Job tracker coordinates the job on basis of their processing speed, number, and time via MapReduce. Data node reports to name node about block information of data. Apart from this, data node also sends its active report to name node by sending a signal in every 3 s with a message "I am up and I am alive." However, if name node does not get this signal in 3 s, I will consider data node as dead. Task tracker runs on data node and gives reports of status of running tasks to name node. Although HDFS is considered as a robust system, there is a risk of unauthorized access to an HDFS client via RPC or via HTTP [9] (Fig. 1).

Table 3 Overview of MapReduce

Hadoop MapReduce	
Developed by	Google
Developed for	Batch processing
Used language	Java language
Output storage	Hard disk
Internal storage processing reliability	Not reliable
Fault tolerance	By data replication (by default, three replications)
Drawback	Frequent disk input/output usage

4 Key Components of Hadoop Ecosystem

4.1 MapReduce

MapReduce programming is designed for computer clusters [10]. MapReduce is the main component of Hadoop, which is used to process large sets of data on commodity clusters. MapReduce is a processing framework. Programs of MapReduce can be written in any languages such as Java, C++, Python because it supports Hadoop streaming (Table 3).

MapReduce can process any data type via line offset, either it is structured or unstructured data of HDFS, e.g., word count. MapReduce solves the bottleneck issue of traditional data processing systems for storing, analyzing, and processing masses of data in a single system. Major facet of MapReduce is its parallel processing. This results in MapReduce faster execution system.

4.1.1 Tasks of MapReduce are Classified in Two Ways

Map gets inputs in the form of data sets, files, or directories stored at HDFS. Input files are passed to mapper linewise, and it splits the data sets into various individual tuples and generates output as key–value pair. Reduce takes the output of map as its inputs, does the work of logical combining of tuples, and stores the processed result in HDFS.

4.1.2 Workflow of MapReduce

Client gives its input to job tracker; job tracker connects with name node and searches the client's requested data; task tracker processes the input as per the instruction of client and gives its present working status to name node. Job trackers fetch the information from name node. Further, the output of the client is again saved to HDFS.

4.2 Apache Pig

Apache Pig is a high-level scripting language, developed by Apache Foundation. Pig uses ETL tool, i.e., Extract, Transform, and Load tool.

Pig provides a platform in Hadoop to customize, analyze, and manipulate large sets of data. Pig language is known as pig Latin. Pig Latin consists of several operations, which if used allows programmer to develop their own functions like reading, writing, processing, etc. Pig Latin lets programmers to write scripts which are then internally converted into the MapReduce task. Pig engine (an Apache Pig component) further takes inputs in the form of these pig Latin scripts and produces output into MapReduce jobs. This output gets stored into Hadoop clusters.

Pig accomplishes pig Latin through grunt shell. Grunt shell is used to write scripts in pig Latin language by invoking its commands. Two commonly used commands in grunt shell are "sh" and "fs". Grunt shell also gives utility commands like clear, quit. Pig supports Hadoop streaming, and it can accept program written in any languages like Java, Python. Pig inherits MapReduce framework to process data. Pig was actually developed for non-Java programmers in order to make it efficacious for every programmer.

4.2.1 Data Types in Pig

In Fig. 2, *tuple* acts as row in which records are mentioned in an ordered form into fields of any type. *Bag* acts as a table and is represented by "{ }". Every bag has individual number of tuples. *Map* contains key–value pairs, where identity of key must be in unique and in character type and value can be of any type.

Map is represented by "[]". *Atom* is a small part of data stored in string. Pig also supports user-defined functions which featured it as extensible and allows programmers to make their own data types in "bin" folder of pig Latin scripts.

Data Types in Pig		Implemented Classes as per Java
Complex Data Types	Bag	org.apache.pig.data.DataBag
	Tuple	org.apache.pig.data.Tuple
	Map	java.util.Map<Object, object>
Scalar Data Types	Integer	java.lang.Integer
	Long	java.lang.Long
	Float	java.lang.Float
	Double	java.lang.Double
	Chararray	java.lang.String
	Bytearray	byte[]

Fig. 2 Data types usage in Apache Pig

4.2.2 Workflow in Pig

Programmer writes their scripts using pig Latin language along with their supported execution mechanism (e.g., grunt shell, user-defined functions). After successful execution, scripts go for a series of transformation that includes compiling and optimizing the scripts, and then, internally, these scripts get converted into MapReduce scripts. Further, these scripts are forwarded to MapReduce framework and then saved or written to HDFS.

4.3 Hive

Hive is data warehousing software used for processing of structured data. Hive was developed by Facebook, but later, Apache Software Foundation took it up from Facebook and released it as open-source software with the name "Apache Hive." Hive uses Hive Query Language (HQL) similar to SQL. Hive is highly used in Hadoop ecosystem for writing queries and developing applications of Hadoop. When it comes to process structured data, Hive is generally more reliable than all others. Hive basically supports three kinds of data types: integral data type, literal data type, and string data type.

4.3.1 Terminologies Related to Hive

Hive user interface: Hive is data warehouse open-source Apache software that allows users to interact with HDFS. Hive-supported user interfaces are Web user interface, Hive command line, and HD insight.

Meta store: Hive has its own database servers to store table's metadata, their data type, and mapping. These servers are known as Meta stores.

HQL process engine: HQL is similar to SQL for querying data in Meta store. In spite of writing MapReduce programs with traditional approach, it is better to write a query for MapReduce job and further process it.

Execution engine: It works as junction between HQL process engine and MapReduce framework. It works the same as MapReduce.

HDFS or HBase: HDFS or HBase are data storage repository to store data.

4.3.2 Workflow in Hive

First of all, Hive interface like command line sends query of data to drivers for accomplishment. Driver checks the syntax and query process with the help of compiler, and then, the compiler sends a request for metadata to Meta store. Here, query gets compiled. Driver again sends the executed plan to execution engine. Internally, process execution is jobs of MapReduce. Execution engine sends the data as job to job tracker under name node. Here, query is accomplished as MapReduce job. At last, execution engine fetches output from data node and transfers it to driver and driver shows output at Hive interface.

4.4 Sqoop

Sqoop is the combination of SQL and Hadoop. Sqoop acts as a data transfer bridge between Hadoop and relational database servers such as SQL. Sqoop main work is to import and export data. Sqoop works as subtool in Hadoop modules for processing data. Sqoop Meta store works as a storage system that stores data being imported to Sqoop and processed outputs that need to be transferred to centralized systems. It simulates multiple tasks to be done in the meantime. Sqoop Meta store also works as incremental loader that holds the last updated value of transaction of data.

4.4.1 Workflow of Sqoop

Sqoop extracts data in the form of tuples and bags from relational databases like SQL and imports it to HDFS. Each tuple in bag is then transformed as records in HDFS and stored as text files. Further, these text files are exported to Hadoop file system (HDFS, Hive, and HBase).

4.5 Apache Flume

Apache Flume works as a data management tool for streaming data from several sources to centralized data store (let HDFS). Flume works in distributed environment with high reliability and fault-tolerant ability. Nowadays, flumes like services are highly used in IT sectors for data safety, record keeping, and faster transfer of data to data storage servers. Flume is capable of fetching log data and events from multiple Web servers into a centralized database storage. Flume acts as a mediator between Web servers and database storage software and provides a steady exportation of data between them. It keeps track of data transfer rate. If in any case, data transfer gets higher than the data written rate in database server, flume acts as a controller too. Flume assures accurate content delivery from source to destination address with contextual routing.

4.5.1 Terminologies Related to Flume

Log file: Log file is a data storage that stores generated actions on current processing. Flume agents: Flume has agents that internally acts as a Java Virtual Machine process and contains commands by which events get transferred to next destination.

4.5.2 Workflow in Apache Flume

Web servers such as Facebook, Amazon, Flipkart generate log data in tremendous amount. These data are then collected by flume agents that are connected with flume service. Entire data gets collected from flume agents and gets customized. Customized data is then transferred or written in centralized stores such as HDFS or HBase.

5 Preference of Hadoop Technology over Traditional Database

Traditional database systems (e.g., RDBMS) consist of ACID properties: Atomicity, Consistency, Isolation, and Durability. But when we talk about Hadoop, we must understand first that it is not a database system but it contains similar functions such as extracting, manipulating, storing data like RDBMS; however, the terms of data processing in both the methods are different. Hadoop basically works with its two components: HDFS (storage system) and MapReduce (retrieves data from Hadoop clusters). Both RDBMS and Hadoop work for processing data only, but RDBMS can only process well-structured data in tuples with particular schemas based on ER models. Example of RDMS is online transaction processing (OLTP). RDMS now becomes unreliable with the pace of time because it cannot deliver fast results and needs more CPU storage. Hadoop system precisely manages all types of data formats with high fault tolerance capability by its clusters. Hadoop do deliver faster execution result which is the need of today's world.

One cannot manage data now without its proper storage functions that happens in traditional database management systems. Hadoop is the key to this problem. Database systems are built for multi-step transactions and high power statistics apart from basic data. In the present era, these complicated systems are inefficient for extracting and processing bulk amount (in 100s of terabytes). Hadoop is meant for storing this bulk data at massive speed. Hadoop is developed for allocation of information systems that possess point-to-point details with inconsistency with respect to time.

6 Summary

Hadoop is a software framework that can be installed on a commodity Linux cluster to permit large-scale distributed data analysis [11]. Nowadays, organizations release tremendous amount of data every day. Hence, database administrators (DBAs) have toughest job of maintaining crucial data with proper security. Any database administrator (DBA), who is working with traditional database, will get resultant of certain disadvantages (few are no room for unstructured data, no real-time analysis, etc.). These drawbacks pushed back the organizations from reality of evolution. For example, Amazon gets real-time analysis of their consumer's feedback with their approximately 232 billion products. Hadoop is an open-source software platform for distributed computing dealing with a parallel processing of large data sets. It has been widely used in the field of cloud computing [12]. Hadoop is a framework that inherits distributed processing of large data sets across clusters of commodity computers using a simple programming model that can also tolerate fault and automated system failure. The volume and the heterogeneity of data with the speed it is generated make it difficult for the present computing infrastructure to manage Big data [13]. When it comes to cost, Hadoop is cheaper because of its clusters than traditional database systems. DBA professionals should move to Hadoop on both organizational level and individual level. Hadoop is on current trend according to the Big data Executive Survey of 2013 which states that "almost 90% organizations have implied Hadoop-related projects on their ground level." The MapReduce paradigm has emerged as a highly successful programming model for large-scale data-intensive computing applications [14]. MapReduce is a parallel processing system that works on distributed commodity clusters rather than serially which definitely saves time. MapReduce is a programming model and an associated implementation for processing and generating large data sets [15]. Suppose Amazon wants to calculate its yearly sales city-wise. Amazon has 1 terabyte of data on traditional processing system. As a result, with billions of products, this amount of data space will run out of memory. Hence, Amazon uses MapReduce. In MapReduce, there are two phases: Map and Reduce. Here, rather than giving complete job to one phase, Amazon splits whole data into small chunks on the basis of maps. These mappers work parallel to fractional data. After the completion of mapper's task, Reduce phase takes work on their area by fetching outputs of mappers (intermediate records) as their inputted data, sorts them if needed, and further gives output as needed. Basically, reducer reduces a set of intermediate values which share a key to a smaller set of values. Somebody who is working with traditional database like SQL will look at Hadoop like a big mess. Main criteria stand here are for handling supported data types. Traditional database systems cannot handle unstructured and semi-structured data, whereas Hadoop is capable of handling all kinds of data with sophistication.

References

1. Bagriyanik S, Karahoca A (2016) Big Data in software engineering: a systematic literature review. Glob J Inf Technol 6(1):107–116
2. Tsai CW, Lai CF, Chao1 HC, Vasilakos AV (2015) Big Data analytics: a survey, of Big Data 2:21. https://doi.org/10.1186/s40537-015-0030-3
3. Saltz JS, Shamshurin I (2016) Big Data team process methodologies: a literature review and the identification of key factors for a project's success. In: 2016 IEEE International Conference on Big Data (Big Data)
4. Nelson B, Olovsson T Security and privacy for Big Data: a systematic literature review. In: 2016 IEEE International Conference on Big Data (Big Data)
5. Kumari S A review paper on Big Data and Hadoop. Int J Recent Adv Eng Technol (IJRAET) 4(1):2347–2812 (For National Conference on Recent Innovations in Science, Technology & Management (NCRISTM) ISSN (Online))
6. Ularu EG, Puican FC, Apostu A, Velicanu M (2012) Perspectives on Big Data and Big Data analytics. Database Sys J III(4)
7. Anjali PP, Binu A (2014) A comparative survey based on processing network traffic data using Hadoop Pig and typical map-reduce. Int J Comput Sci Eng Surv (IJCSES) 5(1)
8. Assunção MD, Calheiros RN, Bianchi S, Netto MA, Buyya R (2015) Big Data computing and clouds: trends and future directions. J Parallel Distrib Comput 79–80:3–15 (Elsevier)
9. Mukherjee S, Shaw R Big Data—concepts, applications, challenges and future scope. Int J Adv Res Comput Commun Eng 5(2)
10. Sreedhar C, Kasiviswanath N, Reddy PC (2017) Clustering large datasets using K-means modified inter and intra clustering (KMI2C) in Hadoop. J Big Data (Springer)
11. Taylor R (2010) An overview of the Hadoop/MapReduce/HBase framework and its current applications in bioinformatics Author. In: Pacific Northwest National Laboratory Bioinformatics Open Source Conference 2010 Richland, WA
12. Lu H, Hai-Shan C, Ting-Ting H (2012) Research on Hadoop cloud computing model and its applications. In: 2012 third international conference on networking and distributed computing
13. Dhavapriya M, Yasodha N (2016) Big data analytics: challenges and solutions using Hadoop, map reduce and big table. Int J Comput Sci Trends Technol (IJCST) 4(1) Jan–Feb 2016
14. Wang L, Taoc J, Ranjan R, Marten H, Streit A, Chene J, Chena D (2013) G-Hadoop: MapReduce across distributed data centers for data-intensive computing. Future Gener Comput Sys 29:739–750, Elsevier
15. Dean J, Ghemawat S (2004) MapReduce: simplifed data processing on large clusters. research.google.com/archive/mapreduce

Smart Mobile Bot Detection Through Behavioral Analysis

Iroshan Aberathne and Chamila Walgampaya

Abstract Mobile advertising became a huge financial pillar due to drastic increase in smartphones and tablets usage in recent years. This huge-revenue ecosystem is severely thwarted by ad fraud due to large sum of money available in this market. Trained botnets and even individuals are hired by click-fraud specialists in order to maximize the revenue of certain users from the ads they publish on their Web sites or to launch an attack between competing businesses. This study proposes a novel and far efficient real-time approach to identify and categorize real mobile users over click bots through behavioral analysis. To validate the effectiveness of our approach, Real Time Mobile Bot Miner (RTMBM), an architecture based on the proposed methodologies has been implemented. The concept behind the RTMBM is how a user reacts to unexpected/dynamic User Interface (UI) changes in a Web page. Experimental results show two unique behavioral patterns of a real mobile user after an unexpected UI change occurred. The results can easily be adapted to any existing Web site to differentiate a given user from a click bot. This could even be more convenient than identifying captchas, filling text, etc.

Keywords Mobile click fraud · Click bot · Behavioral analysis

1 Introduction

Mobile advertisement market is being expanded in higher velocity due to the rapidly increasing of smartphones and tablets usage in recent years [1]. The main contributors of this market are *advertisers, publishers*, and *advertising networks (ad networks)*. Advertiser is the one who makes a contract with ad network to publish advertisements

I. Aberathne (✉) · C. Walgampaya
Faculty of Engineering, University of Peradeniya, Peradeniya, Sri Lanka
e-mail: aurora.bcg@gmail.com

C. Walgampaya
e-mail: ckw@pdn.ac.lk

© Springer Nature Singapore Pte Ltd. 2018
M. L. Kolhe et al. (eds.), *Advances in Data and Information Sciences*, Lecture Notes in Networks and Systems 38, https://doi.org/10.1007/978-981-10-8360-0_23

on behalf of himself or a company. Publisher can be a Web site which displays the advertisements to the site visitors.Ad network plays a broker role between advertiser and publisher. Advertiser is charged by the ad networks for publishing their advertisements. Ad networks find the suitable publishers to display the ads [2]. By 2017, analysts predict that revenue from mobile advertising will exceed that of TV advertisements and account for one in three dollars spent on advertising [3].The global advertising market is projected to reach over $100 billion and account for more than half of all digital advertising spending in 2016 [4].

However, this mobile advertising market is frequently thwarted by numerous types of frauds [5]. The fundamental fraud type of mobile advertising market is called *"Click Fraud."* Click fraud is the practice of deceptively clicking on advertisements with the intention of either increasing third-party person revenues or exhausting an advertiser's budget [6]. In February 2016, Check Point researchers first discovered HummingBad, a malware that established a persistent rootkit on Android devices. Currently, more than 85 million mobile devices are being controlled by them and generate over $300,000 fraudulent ad revenue per month [7].

The click frauds in mobile advertising generally fall under two main categories called *Bot-driven fraud* and *Placement fraud.*

1.1 Bot-driven Fraud

The concept of a Botnet evolved in 1993 by introducing the first Botnet called Eggdrop [8]. The bot is an intelligent program that operates automatically as an agent for different goals. The bot term denotes to an infected computer by malicious code which often exploits software vulnerabilities on the computer to allow a malicious party to control the computer from a remote location [8].

Figure 1 illustrates a visual layout of botnet. A Botnet is a network of infected computers under the remote command and control (C&C server) of an operator called Botmaster [8]. In the mobile platform, MoBots or Mobile botnet have become the most critical challenge in mobile click fraud.

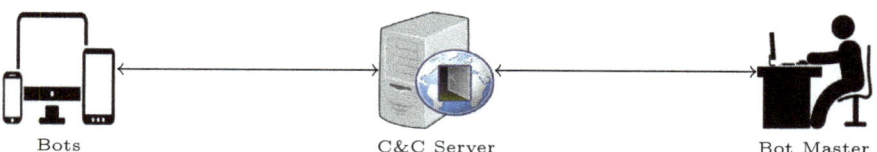

Bots   C&C Server   Bot Master

Fig. 1 Bot net

1.2 Placement Fraud

In general, this can be done by manipulating visual layouts of ads, including resizing ads to too small to read, hiding them behind other display element such as UI buttons or placing them outside the display screen to trigger ad impressions and unintentional clicks from real users [5].

Differentiating whether a click or a touch is from a real user but not from a bot is obviously very difficult in current mobile platform [3]. Although network advertisers and publishers seek to defend themselves in many different ways, fraudsters have become more sophisticated, using increasingly complex programs and scripts which are able to spoof the source of the clicks and perform the fraud automatically [9].

The majority of researchers have proposed varies kind of approaches to investigate and track click fraud. Some of them are data analyzing techniques [10, 11], detection tools [12], cryptographic approaches [13, 14], and algorithmic approaches [15, 16].

The challenges faced by existing mobile bot detection and prevention solutions have given insight to our proposed solution. In this paper, we proposed a novel and far efficient bot detection technique in real time for mobile devices. The remainder of this paper is organized as follows. We examine related work in Sect. 2. Then, we detail our approach in Sect. 3 and validate its efficiency with real data in Sect. 4. Finally, we conclude the paper in Sect. 5.

2 Related Works

Crussell et al. [17] proposed solutions called MAdFraud which observe abnormal/ deviated behaviors of mobile ad fraud prepared by Android apps. It identifies two kinds of fraudulent behavior in controlled environment. They are requesting ads in the background and clicking on ads without user interaction. There are number of drawbacks of their approach. First, they run apps in emulators but not in real devices. Some ad libraries do not display ads in emulators. Second, they cannot detect placement fraud. Because MadFraud detects fraud that is performed silently to the user, either ads requested in the background or ad clicks without user interaction. However, their study may not be sufficient to show the real impact of click-fraud attacks in mobile platforms because their study mainly focused on analyzing existing mobile applications' fraudulent behaviors that could be used for advertisement fraud.

DECAF [5] characterizes click fraud in mobile apps and proposes offline techniques to detect display on Windows-based mobile platforms by using automated testing. Their detection system is bounded to placement fraud which means analyzing the status of ad UI offline to determine whether the ads are hidden, obfuscated, or stacked. Their approach totally depends on rule-based method to identify fraud apps/pages. With this rule-based method, DECAF fails to detect smart bots fraud activities. DECAF focuses on user-based ad fraud in the mobile apps setting rather than the click-spam fraud in the browser setting.

Cho et al. [18] implemented a test app called ClickDroid that automatically generates click events for advertisements on selected ad networks. Under the pay-per-click model, both ad network and publisher are beneficial while ad advertiser is the only victim [19]. So it is better to have prevention or protection techniques on advertiser side rather than on the ad network side. Their focus area is ad network rather than advertiser. Anyhow these kind of apps can be simply be detected by pattern recognition mechanisms. Oentrary et al. [20] and Kitts et al. [11] introduced systems to detect such fraud patterns in server-side. The researchers have only focused on Android mobile environment.

AdAttester [3] tries to detect and prevent well-known ad fraud by identifying incoming click or impression which is actually delivered by a real user with two primitives called *verifiable display* and *unforgeable clicks*. AdAttester uses a third-party software called ARM trustZone technology for user identification and device identification. Keeping such dependencies on third-party software is not suited for this kind of fraud detection tools. Apart from that to deploy AdAttester, mobile user needs to install attestation application as well. AdAttester cannot detect mobile ads that violate the ad policy of the ad provider.

3 Proposed Approach

In our approach, we try to differentiate mobile click bot behavior over real mobile user behavior in real time. Since our way of detecting mobile bot is through behavioral analysis, *Real Time Mobile Bot Miner* (RTMBM) was developed. RTMBM listens to each and every user event within a user session in real time. Figure 2 illustrates the architectural diagram of RTMBM. Mainly RTMBM consists of three layers.

Front-end data collection layer is responsible for collecting all the mobile/desktop generated events and filtering them. Back-end data processing layer is responsible for processing all the incoming data and formulation of more attributes to next layer called decision-making layer. All the decisions are made in this layer at real time.

3.1 User Session

User session is a system defined time period with unique identifier called *session_Id*. All user activities are bounded to the user session. The status of a given session can either be *Active* or *Inactive*. Two algorithms have been developed to maintain and manage the session status. Active user sessions are identified by *Heart-beat algorithm* while session status change is managed by *Scheduler algorithm*.

Figure 3 illustrates both heart-beat algorithm and scheduler algorithm. Both algorithms run in pre-defined time interval (heart-beat pre-defined time interval is less than that of scheduler) by the system.

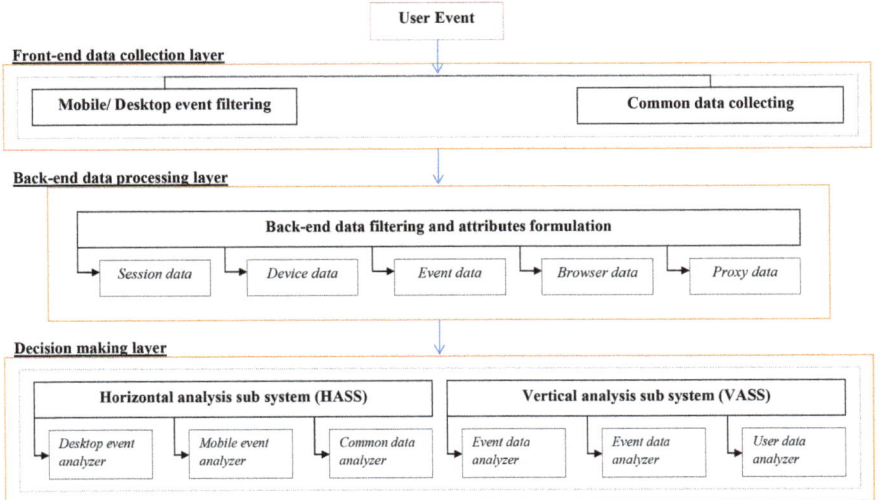

Fig. 2 System architecture

Fig. 3 Types of algorithms

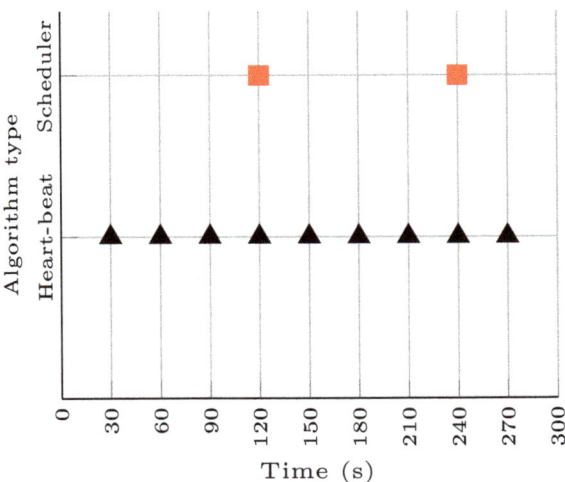

Heart-beat algorithm: *Heart-beat algorithm* is used to send notification requests (heart-beats) to the server regularly in pre-defined time interval. A given active session executes this algorithm to send heart-beats to the server indicating that the session is active. Then RTMBM keeps latest heart-beat time stamp in the database against *session_Id*.

Scheduler algorithm: *Scheduler algorithm* also runs in pre-defined time interval to change the session status depending on latest heart-beat time stamp. If there is no heart-beat time stamp during past 3 min (or system defined time), scheduler changes the session status to "INACTIVE."

Fig. 4 User session status

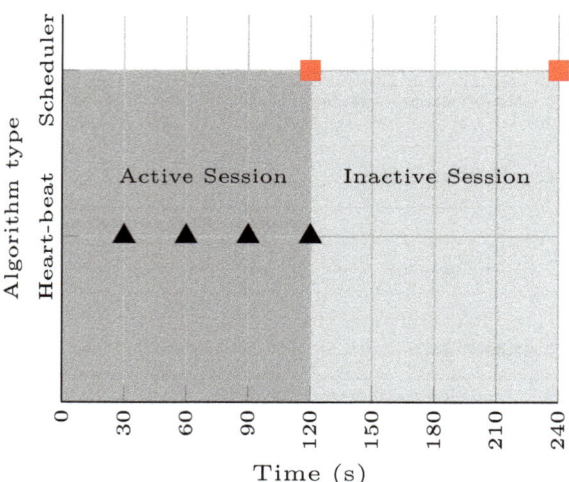

Figure 4 illustrates the user-status changing process with heart-beat and scheduler.

3.2 User Event

User event is any kind of user interaction with the system within a user session. The user events can be either mobile events or desktop events. As an example mobile event can be touch start, touch move, swap, touch zoom, etc. Desktop event can be left click, right click, scroll, etc.

3.3 Front-end Data Collection Layer

There are two subsystems in the front-end data collection layer to collect and filter incoming traffic.

Common data collecting subsystem: collects data which are common to both desktop and mobile devices. Common data can be further subgrouped into three categories.

1. Event identification common data:

   - Event triggered time, event triggered position coordinate, offset of the html tag and event triggered html tag name, etc.

2. Device identification common data:

   - Device screen width/ height, device orientation, device viewport width/ height, etc.

3. Location identification common data:

- Time zone offset and device time. For example, for traffic generated from the US time zone offset will be from −10:00 GMT to −04:00 GMT. In Sri Lanka, it will be +05:30 GMT

Desktop/Mobile event filtering subsystem: Most of the mobile events triggers chain of subsequent mobile and desktop events in background. As an example touch start event automatically triggers set of mobile events such as touch move, touch end and then triggers desktop events like mouse over, mouse move, mouse down, mouse up, and eventually mouse click.

We have developed separate subsystems (algorithms) using advanced jQuery and Spring mobile technologies to address this generic behavior of the mobile events. Implementing this algorithm RTMBM clearly detects triggered events.

Then all gathered data are validated at the front-end layer and then forwarded to back-end data processing layer.

3.4   Back-end Data Processing Layer

As Fig. 2 illustrates, there are five back-end data processing and attribute formulation processes.

Session data: The basic building block of the RTMBM is *session*. RTMBM maintains six session-related attributes such as session_Id, session created time, session last access time, session access count, session heart-beat time, and session status.

Device data: Most of the device-specific data such as device width/ height, viewport width/ height, operating system, device type (mobile or desktop), and orientation are filtered from incoming front-end data.

Event data: Number of data collected at front end such as event type (touch/ click, zoom, etc.), event triggered time (device time), event-triggered position coordinate, number of taps, event-triggered html tag name, and image name are filtered here. Apart from that, time zone of that country and actual time of that country are derived from time zone offset attribute with another back-end process.

Browser data: Browser-related data such as browser name, browser type, browser version, browser id, and referrer details are collected from user agent string.

Proxy data: Separate database is being used to retrieve proxy related data such as number of proxies that user has been used, IP address, country name, city name, postal code, latitude and longitude of each proxy.

3.5 Decision-Making Layer

RTMBM uses two subsystems called horizontal analysis subsystem and vertical analysis subsystem to make decisions about if a user is genuine or not.

Horizontal analysis (User event analysis): In the horizontal analysis, we developed three different analysis models called Desktop event analyzer, Mobile event analyzer, and Common data analyzer. It is identified that there are relationships exist among attributes in single user event.

Vertical analysis (User session analysis): In the vertical analysis, we use all the user events of a session to detect anomalies in a single session. RTMBM uses Event data analyzer, Device data analyzer, and User data analyzer to analysis session events.

3.6 Extended RTMBM to Trigger Unexpected User Interface (UI) Changes

In our approach, we try to differentiate mobile click bot behavior over real mobile user behavior in real time. Since our way of detecting mobile click bot is behavioral analysis, RTMBM listens each and every user event within a user session in real time.

The concept behind the extended version of RTMBM is how the user reacts to unexpected/ dynamic UI changes in a particular Web page. There can be number of unexpected UI changes when user browses a Web page. As an example page loading time, image loading time and data retrieving time can be higher due to low bandwidth Internet connection.

Extended version of RTMBM is capable of artificially generating these kind of unexpected/dynamic UI changes in a Web page. To generate unexpected/dynamic UI changes, RTMBM has two options.

Automatic: If user session behaves abnormally and reaches to a threshold value, RTMBM automatically triggers UI changes to Web page and listens to the user events.

Manual: Admin user (Logged in user of the RTMBM back end) can manually triggers the unexpected UI changes to a Web page.

The current version of RTMBM is capable of disabling all the cascading style sheet (CSS) of a Web page to make the unexpected UI changes. Figure 5 illustrates actual screenshots of a Web page before and after CSS is disabled. RTMBM collects all user events to further processing after unexpected UI changes occurred in a specific user session either manually or automatically.

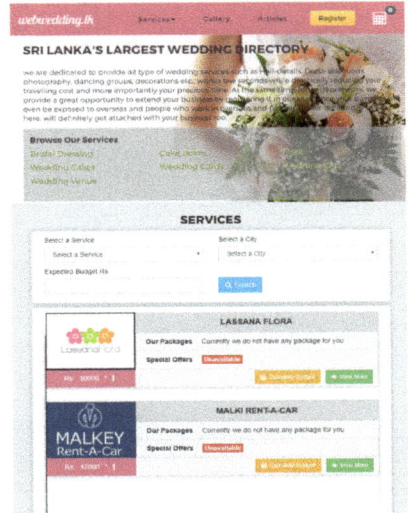

Fig. 5 Web page with CSS and without CSS

3.7 Survey Scope and Boundaries

To evaluate our approach, we conducted a survey with beta version of RTMBM to collect real-world data. The beta version of RTMBM is available online at http://aurorarti.herokuapp.com. All our experimental data are being gathered from http://webwedding.herokuapp.com website.

Since our objective is to identify a behavioral pattern of real mobile users, we have collected data from real human mobile users. Data have been collected from both smartphone users and tablet users. The operating systems were Android, IOS, and Windows. Browsers were Chrome, Firefox, and Mobile Safari. Data have been collected for maximum of three minutes time duration from each user.

4 Experimental Results

Survey data have been gathered from 101 human mobile users. Figure 6 shows the survey data with regard to Refresh (RF) event trigger status. 80 users out of 101 have been triggered RF event when there were any unexpected/dynamic UI changes in a Web page. Remaining 21 (20.80%) users did not trigger RF event even though they noticed the unexpected/dynamic UI changes. As a percentage of 79.20 %, real human mobile users move to trigger RF event while they notice unexpected/dynamic UI changes in a Web page.

As Table 1 depicts 61 human mobile users out of 101 (60.39 %) trigger a RF event within first minute after identifying the unexpected UI changes in a Web page.

Fig. 6 Refresh event triggered status

Table 1 Refresh event triggered users over time

Time Duration (s)	Users	Percentage (%)
0–60	61	60.39
61–90	12	11.90
91–180	7	6.93

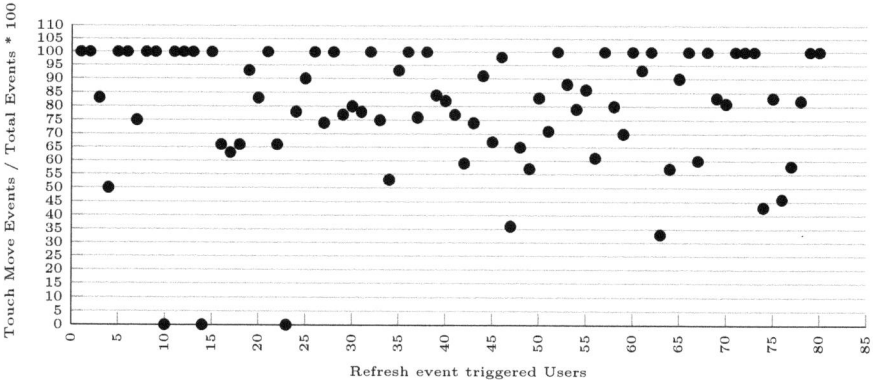

Fig. 7 Touch move events over total events per user

Figure 7 illustrates the Touch move (TM) events/ Total events (TE)-triggered pattern of users during the time period from unexpected UI changes occurred to RF event triggered. Higher proportion of TM events has been triggered over TEs. Only three users did not trigger any TM events. 73 users out of 80 (91.25%) have been triggered more than 50% of TM events over TEs after unexpected/dynamic UI changes occurred.

5   Conclusion

In this paper, we have proposed a novel and far efficient approach to detect mobile click bots over real mobile users. Our proposed detection system performs browsing behavior analysis of a user after unexpected UI changes occurred. Experimental results show two behavioral patterns of a real mobile user. First one is higher proportion (79.20%) of the real mobile users trigger RF event. Apart from that 60.39% real mobile users trigger RF event during first minute. Second behavior is a real mobile user which tends to trigger more TM events (91.25%) over other events such as Touch start or Zoom events before triggering RF event. The RTMBM can easily be adapted to any existing Web site to differentiate a given user from a click bot. The proposed method can be used to replace or substitute the use of captchas, filling text, etc.

References

1. Cho G, Cho J, Song Y, Choi D, Kim H (2016) Combating online fraud attacks in mobile-based advertising. EURASIP J Inf Secur 2016(1):1
2. Alrwais SA, Gerber A, Dunn CW, Spatscheck O, Gupta M, Osterweil E (2012) Dissecting ghost clicks: Ad fraud via misdirected human clicks. In: Proceedings of the 28th annual computer security applications conference. ACM, pp 21–30
3. Li W, Li H, Chen H, Xia Y (2015) Adattester: secure online mobile advertisement attestation using trustzone. In: Proceedings of the 13th annual international conference on mobile systems, applications, and services. ACM, pp 75–88
4. Grewal D, Bart Y, Spann M, Zubcsek PP (2016) Mobile advertising: a framework and research agenda. J Mach Learn Res 15(1):99–140
5. Liu B, Nath S, Govindan R, Liu J (2014) DECAF: detecting and characterizing ad fraud in mobile apps. In: 11th USENIX symposium on networked systems design and implementation (NSDI 14), pp 57–70
6. Wilbur KC, Zhu Y (2009) Click fraud. Mark Sci 28(2):293–308
7. HummingBad Research report FINAL 62916
8. Amini P, Araghizadeh MA, Azmi R (2015) A survey on botnet: classification, detection and defense. In: 2015 International electronics symposium (IES). IEEE, pp 233–238
9. Costa RA, de Queiroz RJ, Cavalcanti ER (2012) A proposal to prevent click-fraud using clickable CAPTCHAs. In: 2012 IEEE sixth international conference on software security and reliability companion (SERE-C). IEEE, pp. 62–67
10. Mann CC (2006) How click fraud could swallow the internet. Wired Mag 17–20
11. Kitts B, Zhang JY, Wu G, Brandi W, Beasley J, Morrill K, Ettedgui J, Siddhartha S, Yuan H, Gao F, Azo P (2015) Click fraud detection: adversarial pattern recognition over 5 years at microsoft. Real world data mining applications. Springer International Publishing, Berlin, pp 181–201
12. Haddadi H (2010) Fighting online click-fraud using bluff ads. ACM SIGCOMM Comput Commun Rev 40(2):21–25
13. Juels A, Stamm S, Jakobsson M (2007) Combating click fraud via premium clicks. In: USENIX security, vol 70
14. Blundo C, Cimato S (2002) SAWM: a tool for secure and authenticated web metering. In: Proceedings of the 14th international conference on software engineering and knowledge engineering. ACM, pp 641–648

15. Metwally A, Agrawal D, El Abbadi A (2007) Detectives: detecting coalition hit inflation attacks in advertising networks streams. In: Proceedings of the 16th international conference on World Wide Web. ACM, pp 241–250
16. Immorlica N, Jain K, Mahdian M, Talwar K (2005) Click fraud resistant methods for learning click-through rates. In: International workshop on internet and network economics. Springer, Berlin, pp 34–45
17. Crussell J, Stevens R, Chen H (2014) Madfraud: investigating ad fraud in android applications. In: Proceedings of the 12th annual international conference on mobile systems, applications, and services. ACM, pp 123–134
18. Cho G, Cho J, Song Y, Kim H (2015) An empirical study of click fraud in mobile advertising networks. In: 2015 10th International conference on availability, reliability and security (ARES). IEEE, pp 382–388
19. Xu H, Liu D, Koehl A, Wang H, Stavrou A (2014) Click fraud detection on the advertiser side. European symposium on research in computer security. Springer International Publishing, Berlin, pp 419–438
20. Oentaryo RJ, Lim EP, Finegold M, Lo D, Zhu F, Phua C, Cheu EY, Yap GE, Sim K, Nguyen MN, Perera KS (2014) Detecting click fraud in online advertising: a data mining approach. J Mach Learn Res 15(1):99–140

Compendium Depiction on the Applications of Cloud Robotics for the Reclamation of Mankind

Rajesh Doriya and Kaushlendra Sharma

Abstract Cloud computing in the present time is considered to be one of the most pronounced, protruding, and emerging area of computer science. Cloud robotics basically means the robots with their heads in the cloud; the article streamlines the topic and tries to deliberate the area where cloud robotics is deployed and contributing a noteworthy part in flourishing the respective field like cloud medical robots, industrial robots, programmed robots, Google's self-driving cars, application of heterogeneous robots in a wide geographical area. Counting all these, it inclines to conclude where more it can be diversified to enhance the interest of researchers to pick this technology and contribute a bit more for the betterment of mankind. Cloud robotics is one of the rapidly advancing research areas that allow application programs to deduce the computational time and storage into the cloud. Robots are restricted regarding computational limit, capacity, and memory. Furthermore, cloud gives boundless calculation control, memory, storage, and particularly coordinated effort opportunity. Cloud-empowered robots are distributed into two classes as independent robots and organized robots. The article efforts to accumulate some notable contribution of cloud robotics till date citing an in-depth discussion for the same.

Keywords Medical robots · Industrial robots · Programmed robots
Cloud robotics

1 Introduction

Technology justifies its meaning and properness only when it is served to the mankind or society for its betterment. Robotics in itself is a contrivance to enhance the power of automation; clubbing the concept of cloud computing makes it a prolific combo

R. Doriya · K. Sharma (✉)
National Institute of Technology, Raipur, India
e-mail: kaushlendra84@gmail.com

R. Doriya
e-mail: rajeshdoriya.it@nitrr.ac.in

© Springer Nature Singapore Pte Ltd. 2018
M. L. Kolhe et al. (eds.), *Advances in Data and Information Sciences*, Lecture Notes
in Networks and Systems 38, https://doi.org/10.1007/978-981-10-8360-0_24

to explore more areas and to enrich the feature where already these two individual concepts are contributing. The paper tries to focus on such core areas where cloud robotics has contributed in the past and contributing in the present time. Taking these into consideration, the paper concisely explains the touched area by cloud computing and robotics and further attempts to put forth exploring few more areas where the concepts and features of cloud robotics can be implemented in wide manner making it much more acceptable than before, so that it can be implemented and tested which could enhance the acceptability of this novel concept. The article's primary focus is not only to describe the scientific research inside the laboratory but to convey the exemplary inventions which are helping society and mankind to ease the life. The article also intends to propel few core areas where implementation of cloud robotics can do wonders.

1.1 Cloud Computing

Cloud computing is developed with ages taking advanced shapes from distributed computing and several other networking environments. The core of the cloud computing is sharing resources and virtualization; those terms are collectively called as configurable resources. The high point of cloud computing is that in minimal management, it outperforms when executed with certain application on to the network. From the usage point of view, there are three basic models of cloud computing recognized as Software as a Service (SaaS), Platform as a Service (PaaS), and Infrastructure as a Service (IaaS).

Alongside the cloud computing offers several benefits and advantages for businesses and end users, few of the benefits are self-service provisioning, elasticity, pay per use, workload resilience, and migration flexibility. From deployment point of view, the cloud computing model can be classified into three categories: private cloud, public cloud, and hybrid cloud.

SaaS: Software as a Service is one of the model of cloud computing which is growing rapidly. The reason to that is the ease of access to number of application which can be directly fetched from the Web without actually paying the complete price for that software; one of the best advantages of this is that it facilitates to pay as per the usage. Second prominent advantage of this model is the end user need not to worry about the maintenance of the particular software or application. It is being managed and controlled by a third-party vendor. Most of the application can be directly run from the network without actually downloading and installing to the end nodes.

PaaS: Providing application and software as a service, cloud computing also provides the platform as a service. The developers import advantage of this for development of applications and its modules; this particular service provides the framework upon which the developers can build upon and customize applications.

Fig. 1 Cloud computing model

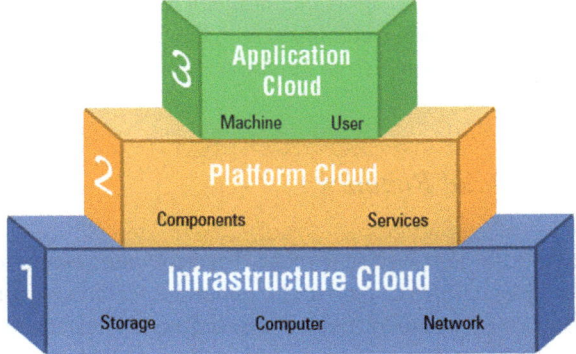

Testing and virtualization also become quite feasible as compared to the conventional techniques. Having all this facility, the developer only has to manage the applications only.

IaaS: IaaS is another form of cloud computing which provides a high degree of virtualization of numerous computing resources (all act as a node) and enables users with the entire requisite infrastructure for computing. Virtualization supports in increasing the usage percentage of resources over the network; the entire facility is installed and placed over the Internet. Scalability is another feature of this model which adds another feather in counting advantages of cloud computing models. IaaS platform carries numerous scalable resources which can accustomed as per the runtime demand of the system. Thus, it makes IaaS very well suited for the jobs which are dynamic in nature and requirements occur abruptly (Fig. 1).

1.2 Robotics

The word robot was first time used in a play by Karel Capek in 1921, where it was showcased as a manufactured slave who can exactly work like human perhaps much better than in terms of accuracy and efficiency. The word robot was derived from a Czech word '*robota*' which roughly means "compulsive servitude." The exact word robotics was first used in 1941 by a science fiction writer, Isaac Asimov, in 1941. And now after a long evolution era, a robot is now defined as a programmable mechanical device that can perform tasks and interact with its environment, without the aid of human interaction. Robotics acts as a bridge between the different fields of engineering and technology which incorporates the concept of mechanical engineering, electrical engineering, computer science, information technology and others for automaton.

Robotics precisely is the technique behind design, manufacturing, and applications of robots. Currently, it is one of the most emerging fields of science and engineering and has contributed in a large way to various fields. The paper has pointed few of the notable fields like in medical science, educational sector, army, agriculture. Robotics when combined with the concepts of cloud computing is getting more happening response from numerous fields.

1.3 Cloud Robotics

Robotics when clubbed with the cloud computing techniques to make it more deployable tries to inherit some of the important features of computing like shared storage, network environment and then converging all the techniques into one. Robotics takes a great advantage of these by sharing the characteristics with all the application programs present in the cloud [1].

Cloud robotics is a new paradigm where robots (application programs) communicate with each other over the Internet. In doing so, it renders a wide reachability and acceptability in many terms like sharing data and code which ultimately results in building emerging research in cloud computing; it also inherits concepts and technology of Big Data and Internet of Things from the surrounding. These technologies help in finding more ways to explore; the only reason for this is that the application programs (more than one) which need to be controlled individually can now be remotely and centrally controlled from one place. That also helps in reducing the space complexity and redundancy of repeated program execution (Fig. 2).

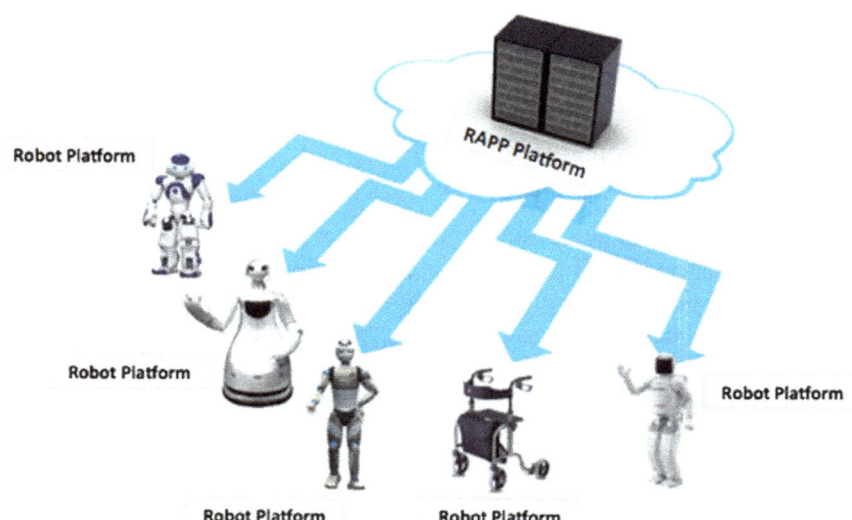

Fig. 2 Prototype of cloud robotics

Table 1 Market growth of cloud robotics

Worldwide cloud robotics market (as per CAGR)	Market revenue in million dollar
Till 2016	2020.66
2016–2022 (expected)	3031.6 (32% growth predicted)

2　Market Growth of Cloud Robotics

The research from one of the renowned organization 'infoholic research' community shows that a tremendous growth has been observed in the field of cloud robotics. Table 1 depicts the growth in the recent years.

3　Application Areas

Industrial Robot: Manufacturing is one of the core areas of development to lead in the competitive scenario. Application of robotics started in the late 1990s in the industry in so many ways [2]; they are technically termed as industrial robots. Continuing the procession, the industrial robots and programmed robots had attained a good level of accuracy and sharp performance in real-time applications in many sections of the industry. Few parameters to judge their working are accuracy, robustness, compatibility with the surrounding, and high-level performances. The mechanical robot is a solid match for some applications [2, 3]. It is regularly utilized for circular segment welding, material taking care of, and gathering applications. They are assembled by number of tomahawks, structure sort, size of work envelope, payload ability, and speed. A robot controller gives the interface to programming and working the modern robot. Below given Fig. 3 depicts the year-wise rapid increase in demand of industrial robots.

　　Introduction of cloud is not new to the industrial robots; it is basically planting a common database for all the robots so that they can be remotely controlled from one place and can be given instruction from a single place [4]. Various combinations of articulated and parallel robots are used in a network in a hybrid manner to work and control different units of industry (Fig. 3) [5].

　　The high-speed demand of robots in the industry has raised the issue of safety and reliability to promise for delivering the expected outcomes. At present, there are so many robots being used in the industry for variety of purposes but the two most widely used robots deployed in majority are articulated robot and parallel robot (Fig. 4).

Medical Robots: Medical robots or surgical robots are one of the most crucial and dignified application of the concepts of cloud robotics which helps the surgeon while handling delicate issues. Another version of these is called as tele-manipulators;

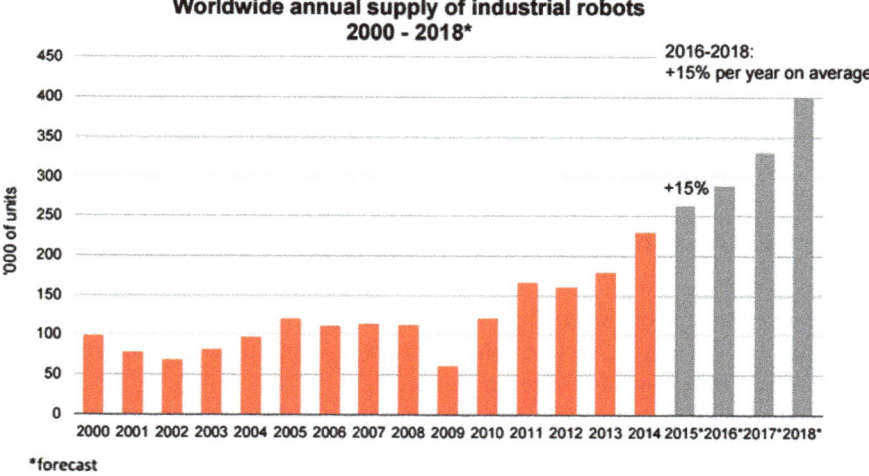

Fig. 3 Annual supply growth of industrial robots

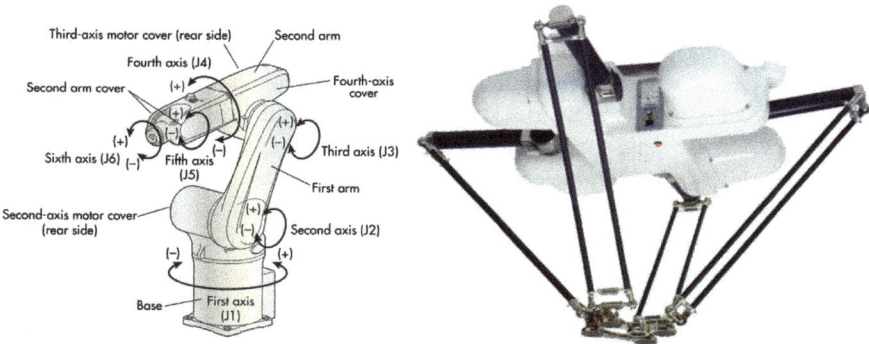

Fig. 4 Articulated robot and parallel robot

rehabilitation robots are another diversified version of robots in medical science. A cluster of robots takes care and enables the lives of needy patients and old-age people, or someone suffering from major ailments. There is also certain cluster of robots which are being used for training and related processes [6].

The advantage of using medical robots can be clearly observed in the surgery; the robots trained for this purpose are much more precise and effective in performing surgeries. Once the robots are well trained and programmed for a specific task, they execute it in a very optimized way by minimizing invasiveness and maximizing biocompatibility [6, 7] (Fig. 5).

Robotics in the uneven Geographical Area: is another most important area where robots are applied at present but can be made more intelligent with the help of cloud and its concepts. Robots basically are used mainly for transport, take care, rescue, and

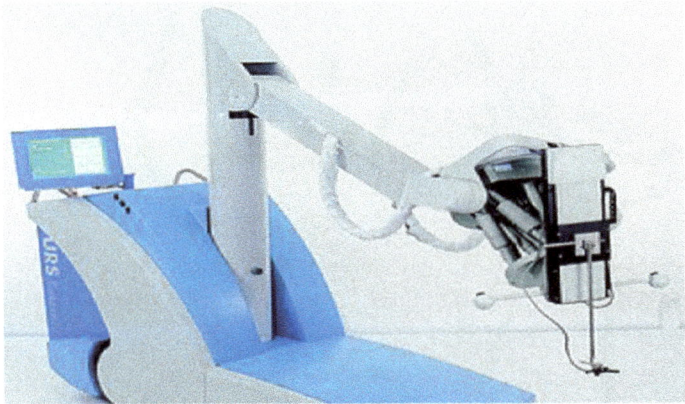

Fig. 5 Model depicting medical robot

attack whenever there is any need at the spot. And with commencement of Big Data and cloud computing with robotics, the scenario has completely taken a new shape while there have been many advances at the equipment level to support the different nodes of battlefield taking it from land to air [8, 9]. It has helped and strengthened the attack and defense by providing synchronization among scattered armed resources. With the growing concepts of robotics, the interoperability and standardization have reached to another level. It has also abandoned the enemy's cunning practices on the battlefield. Network-centric warfare has become the new way to behave on the battlefield.

Robotics in Educational Sector: Many premiere institutes like IIT's/NIT's in India nowadays have implemented effective methods to carry out individual study [10]. There are several concepts and methodologies evolving rapidly to make individual study and classroom study a smarter and intelligent one [11], the concept of robotics contributing in many fashions to conceptualize the need of the system. Currently, the network teaching has overtaken the prime motto of national and international education development; there are several researches done to introduce the integrated structure and model for current education system [10, 12].

With the passing days, robots are becoming an essential predominantly required component in education sector and are considered as a promising aid for teaching and learning in different ways; there were many researches going on to dig out the ways the robots can be planted in pre-primary to high-level school user (Fig. 6).

Agricultural Robots: Agriculture production needs to get doubled in the coming years to cope up with the demand and supply equation. With rising population, it is essential to raise the production of the agriculture. Thus from the human life's aspects, agriculture is one of the important need of life to be dealt with. The advancement in this sector with the commencement of robots has created a buzz to deploy the technology as much as possible. The accuracy of the implementation in the output forces to implant mechanism of robotics in the agriculture sector.

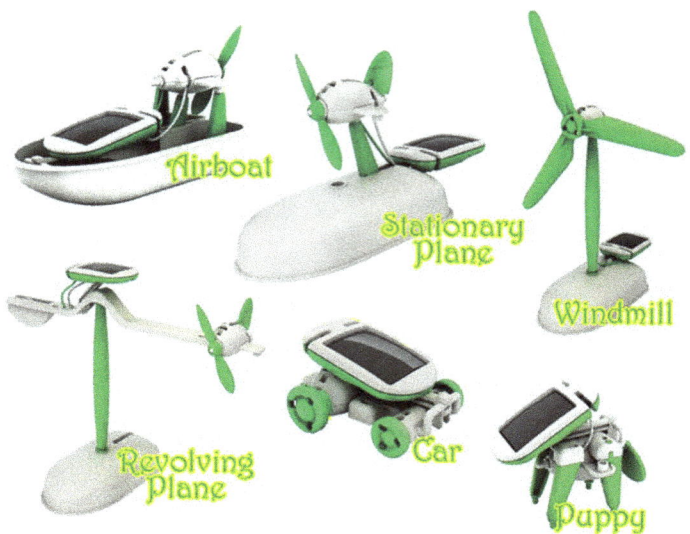

Fig. 6 Prototypes of educational robots

Tele-garden is one among the tested part which has shown significant results. The tele-garden is a community garden [13] that allows users to plant and tend seeds in a remote garden by controlling a robotic arm through a Web-based interface. Such types of robots which are deployed in the farm are also known as Agbot; certain other names are used for them with respect to the work they perform on the field like end effector, gripper, and manipulator. These are deployed for various works like for picking fruits, driverless tractors, and sprinkling. Though, the efforts are being going on to make it a cost-effective [14] [15].

There are various challenges which can be eradicated by using robots in the agriculture field like one of the most important among them is the shortage of labor and to be more confined the biggest problem is of the skilled labor [13, 16]. Fulfilling the above requirement and managing them is a herculean task in the recent times specially in country like India which is mostly dependent on agriculture for its run, facing such problem in spite of having a population of approximately 1.5 billion (Fig. 7).

Fig. 7 Tele-gardening

4 Conclusion

The paper presents a survey report on cloud computing and robotics. The motive of the paper is to present the technical aspects of the cloud robotics but also to show the social impact of that technology for the well being of mankind. Machine (robots) in general is used to keep an eye on human being but with the time changing, the need is to keep an eye on machines also. The concept of robotics when concatenated with cloud computing gives some extraordinary scope to implement cloud robotics on various untouched fields. The paper presents the intense survey on the diversification of cloud robotics implementation. All the advantages and shortcomings are studied thoroughly to conclude how better it can be tested in some other areas. Agriculture sector seems to be more lucrative and useful from various perspectives, agriculture production needs to be encouraged by increasing production, and robots with their heads on the cloud can contribute in a right manner. This is the only field which also reclamates the major human involvement that will help to reduce the problem of unemployment and will grab a great attention from the mass, as it caters one of the major essential need of mankind. There are other major sectors also where robotics is required to contribute for more like need of more surgical robots on medical sectors which can increase the accuracy of sensitive treatments; manufacturing improvement in industries can help to cope up with the increasing demand of commodity, and also industrial automation can't be imagined without Internet robots; educational robots can do wonders, and security of borders in the future will definitely rely on cloud robots.

References

1. http://www.Pinterest.Co.Uk/Haarism/Hnd-Graded-Unit/
2. Corporate ABB, Jiafan C (2017) Challenges in the RAMS database realization for industrial robots. Tcrse
3. Wan J, Tang S, Hua Q, Li D, Liu C, Lloret J (2017) Context-aware cloud robotics for material handling in cognitive industrial internet of things. IEEE Internet Things J 4662(c):1–1
4. Huang C, Zhang L, Liu T, Zhang H (2017) A control middleware for cloud robotics. In: 2016 IEEE international conference on information and automation, IEEE ICIA 2016 Aug, pp 1907–1912
5. Rahimi R et al (2017) An industrial robotics application with cloud computing and high-speed networking. In: Proceedings of 2017 1st IEEE international conference on robotic computing (IRC), 2017, no i, pp 44–51
6. Anwar I, Lee S (2017) High performance stand-alone structured light 3D camera for smart manipulators, pp 192–195
7. Bozcuoglu AK, Beetz M (2017) A cloud service for robotic mental simulations. In: International conference on robotics and automation, pp 2653–2658
8. Sqalli MT et al (2017) Improvement of a tele-presence robot autonomous navigation using SLAM algorithm. In: 2016 international symposium on micro-nanomechatronics and human science (MHS), 2016
9. Wang L, Liu M, Meng MQH (2017) A hierarchical auction-based mechanism for real-time resource allocation in cloud robotic systems. IEEE Trans Cybern 47(2):473–484
10. Cheng Y-W, Sun P-C, Chen N-S (2017) An investigation of the needs on educational robots. In: 2017 IEEE 17th international conference on advanced learning technologies, pp 536–538
11. Sawhney D (2017) Technology integration in Indian schools using a value-stream based framework. In: Proceedings of IEEE region 10 humanitarian technology conference 2016, R10-HTC 2016
12. Jdeed M, Zhevzhyk S, Steinkellner F, Elmenreich W (2017) Spiderino—a low-cost robot for swarm research and educational purposes. In: 2017 13th Workshop on intelligent solutions in embedded systems (WISES), pp 35–39
13. Megalingam RK, Vivek GV, Bandyopadhyay S, Rahi MJ (2017) Robotic arm design, development and control for agriculture applications
14. Wolf D, Prankl J, Vincze M (2016) Enhancing semantic segmentation for robotics: the power of 3D entangled forests. IEEE Robot Autom Lett 1(1):49–56
15. Kahn PH, Friedman B, Alexander IS, Freier NG, Collett SL (2005) The distant gardener: what conversations in the telegarden reveal about human-telerobotic interaction. In: Proceedings of IEEE international workshop on robot and human interactive communication, vol 2005, pp 13–18
16. Jangid N, Sharma B (2017) Cloud computing and robotics for disaster management. In: Proceedings of international conference on intelligent systems, modelling and simulation (ISMS), pp 20–24

A Framework for Data Storage Security in Cloud

Manoj Tyagi, Manish Manoria and Bharat Mishra

Abstract Cloud computing is the growing computing model and highly demandable technology which provide the digital services through Internet. Cloud has huge virtual storage where many users store their data. So selection of efficient and secure server from data centers is important for storing the data to achieve the availability and security. Security of data from unauthorized access is also mandatory. Cryptographic techniques are also used to protect the data by storing it in encrypted form on cloud. Many cryptographic algorithms which are based on attributes, prediction, and identity can be deficient to some level from various security attacks. So there is requirement to develop protected data storage frameworks which ensure the security of data. This proposed work presented security framework for cloud storage, based on metaheuristic approach "modified cuckoo algorithm" for server selection, three-stage authentication, ECC for confidentiality, and CMA-ES for cipher text optimization. The proposed model ensures the availability, authentication, confidentiality, and security of the data while accessing to cloud.

Keywords Cloud computing · Three-stage authentication · Server selection
Modified cuckoo algorithm · Confidentiality · ECC · CMA-ES · Security

1 Introduction

Cloud computing is the pool of data centers which offer resources to clients over the Internet. Example of these recourses is software, computing power, storage space,

M. Tyagi (✉) · B. Mishra
Mahatma Gandhi Chitrakoot Gramodaya Vishwavidyalaya, Chitrakoot, India
e-mail: manojtyagi80.bhopal@gmail.com

B. Mishra
e-mail: bharat.mgcgv@gmail.com

M. Manoria
Sagar Institute of Research Technology and Science, Bhopal, India
e-mail: manishmanoria@gmail.com

© Springer Nature Singapore Pte Ltd. 2018
M. L. Kolhe et al. (eds.), *Advances in Data and Information Sciences*, Lecture Notes in Networks and Systems 38, https://doi.org/10.1007/978-981-10-8360-0_25

etc. Clients may be individual or it may be any organization. Cloud computing gives various type of services which attract every person who wants to use these services in their lives. Cloud computing is rapidly used in many sectors because of its usability, functionality, and characteristics. Cloud computing provides the cost-effective services in which multiple users share the infrastructure, software, and various applications at less cost over the Internet. Here various types of clients pay for only demanded recourses according to their uses. Acceptance of cloud computing in business field is depending on the cost-effective services and trust. So it is the challenge for reliable cloud provider to extend their services and trust in competitive market [1]. Cloud computing offers many benefits to its clients like cost-effective environment, huge storage for storing the data, available all the time for accessing, quick access so that it is adopted at worldwide level and is very popular. Proper separation among various data centers is required in cloud for security purpose. Cloud may suffer from various vulnerabilities, threats, and attacks, and it is the necessity of any cloud to identify and solve these types of security issues [2]. Many individuals and companies use the cloud services, and these clients want assurance from cloud service provider about the safety of their data stored on cloud. Security may be one of the primary reasons that affect the cloud computing growth. So a model or framework is needed that protects the cloud and gives the assurance of security, confidentiality, and privacy of stored data. At present, cloud security has many loose ends that further provide a threat to the confidential data of cloud users. So cloud users are not fully satisfied with the services of cloud technology until there is an approved security framework for cloud. The security framework covers all issue of security by integration of various security approaches in proper order [3]. We also combine the computation and security to achieve the efficiency and confidentiality both. In computation, we focus on efficient and effective resource utilization [4]. Many unauthorized users want to access the cloud services and cause various challenges in data security. So a strong access control is required which permits only authorized user to access the cloud. Access control is the basic and compulsory necessity to prevent the cloud information from unauthorized user [5]. Biologically inspired metaheuristic approaches are used for resource utilization in many fields. So also it is used to manage the cloud servers when users store the data. Bat algorithm, cuckoo algorithm, PSO algorithm, genetic algorithm, firefly algorithm, and many more examples are available to optimize the resources in cloud [6].

2   Related Works

Authentication through password only does not give the assurance that cloud system is fully secure from unauthorized access. Then definitely three-stage-authentication is required for secure access mechanism which uses three factors like CAPTCHA, password, and OTP to verify the client. Three-stage-authentication enhanced the strength of authentication where CAPTCHA and OTP secure the Web site from dictionary attack and phishing attack, respectively [7]. C. Shoba Bindu proposed a click-based graphical (CBG) CAPTCHA in which clients select to sequence of images which are given in CAPTCHA. This CBG scheme provides better result than other CAPTCHA in terms of performance, usability, and security. Suppose CAPTCHA is given in the form of five distorted images in 1×5 grid. Here also display the 15 images in another 3×5 grid. Then clients select the same sequence of images in 3×5 grid which is given in 1×5 grid and accesses the Web page. This CAPTCHA prevents the Web site from dictionary attack and also from spyware attacks [8]. Jiang et al. proposed the OTP scheme that provides less computation as well as high security. It used hash function and SM2 algorithm for this purpose. It prevents from replay attack, impersonation attack, and insider attack [9]. Mahmoudi et al. proposed a MCOA and provide a solution for graph coloring problem [10]. But, we are using MCOA algorithm to select the efficient server in cloud. Singh et al. proposed a generic algorithm for text cryptography applicable to any script with known ASCII value using ECC. Use of ECC has many positive sides like ease of encryption, decryption for large input words, generation smaller cipher text, usage of small key size, in return providing better security than other encryption algorithm [12]. Kampf et al. proposed the hybrid algorithm that uses CMA-ES and HDE optimization for application of solar energy [13]. But in our paper, CMA-ES is used to optimize the cipher text in order to choose the optimal cipher text and enhance the security of cloud. Ahamad et al. in their research work used the mathematical measure thereby improving efficiency of encryption [14].

3 Problem Formulations

Cloud computing has some security concerns with respect to virtualization, storage, and networks. For uploading or downloading the data, cloud permits the clients to access a physical server. But here cloud clients have major concern about the data security. According to the literature, lack of trust, efficiency, and scalability in user authentication and access processes are also challenging issues in cloud-based environments. Clients are authorized to store its data on cloud remotely through online network, and they also access their data when it is required. Because data is stored on remote side, then it is requisite to guarantee that the users' outsourced data cannot be accessed by unauthorized users and is of critical importance to ensure the private information during the users' data access challenges. Many options are available to secure the client's data. But this security was not adequate that is why the proposed integrated framework is employed. This framework blocks various attacks and constructs the system more robust. Here security framework is designed by integrated various individual approaches in this way that it can work in cloud environment efficiently.

4 Proposed System

Cloud computing has high infrastructure, and it is a most recent emergent technique in present trend. Many people's uses the cloud for storing the data so security of these data is very important issue in current scenario. Many good options are available which provide sharing of data in secure manner. But security lacking plays a key challenge for cloud. So a security framework is required to overcome these kinds of security concern. In this paper, we propose a concept of efficient security framework which integrates three-stage-authentication, modified cuckoo optimization algorithm (MCOA) for server selection, elliptic curve cryptography (ECC) for achieving the confidentiality and covariance matrix adaptation evolution strategies (CMA-ES) algorithm for selection of optimized cipher text. Three-stage-authentication permits the authorized clients to access the cloud. In the proposed method, registration phase uses CAPTCHA and OTP where CAPTCHA secure registration process from differential attack and spyware attack. Login phase holds authentication of users using CAPTCHA, clientid, password, and OTP, which carry out user access to get into corresponding user data. OTP prevents the Web site from phishing attack during login. Dynamic server provision is the next stage carried out after the completion of successful user access. For obtain an optimized result in server selection, MCOA is utilized. The designed framework should be prepared to handle client's request for accessing the data after selection of prominent server. After the server's selection, the next phase is secured data access, and this phase paves a path for secured and successful data access to authorized user through a secured framework model. This model encrypts the user data's using any one of the suitable encryption techniques then allows clients to store and access data's in the database. CMA-ES is used to select the optimized cipher text which improves the quality of encryption and more secure the information (Fig. 1).

4.1 Three-Stage-Authentication

A three-stage-authentication protocol first checks the identification of the cloud user and then validates the identity of cloud user through OTP. After authentication, we allow the clients to store their data in encrypted form on cloud server. For storing the data, we are using elliptic curve cryptography for maintaining the security and confidentiality. Here separate server cloud access server (CAS) is used for authentication to improve the security. Four phases are used for this process.

Registration phase: Registration phase provides the platform where any interested client registered with particular cloud. In this phase, CAS provides the clientid and password to client for accessing the cloud services.

Step 1 Clients open the registration form and fill the personal details like name, date of birth, mobile number, e-mail address, security question. CAS also displays the CAPTCHA, and then clients select the CAPTCHA and submit

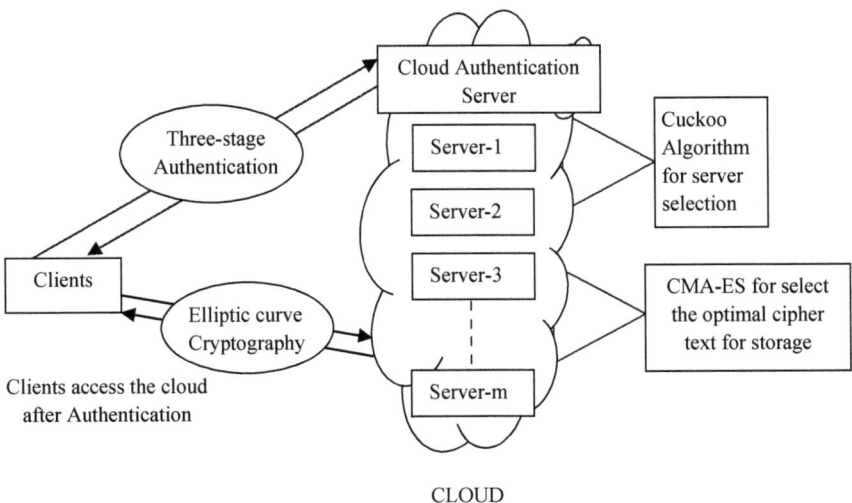

Fig. 1 Proposed framework

the detail. CAS check that selected CAPTCHA is right or not. If CAPTCHA is verified, then clients go in second step.

Step 2 Clients create the clientid and password and then submit clientid and hash value of password to CAS through secure channel. After that, CAS sends the OTP on mobile. Clients enter the OTP in text box and submit its hash value to CAS through secure channel where CAS verifies the OTP, if verified, then registration process is done. Here SH3 hashing is used to secure the process.

Login Phase: Authentication is also done by CAS. Clients enter clientid, password, and OTP in text box then submit clientid, hash value of password and OTP to authentication server. And also select the CAPTCHA based on click-based graphical (CBG) CAPTCHA. After that, CAS receives credential and verifies with stored information. Clients are permitted to access the cloud after authentication.

Password Recovery Phase: In case, if any client forgets the password, then client enters the clientid and request for password. Then CAS sends the temporary password and OTP on registered e-mail and registered mobile number, respectively. After that, client chooses the new password with the help of temporary password, security question, and OTP.

Password Reset Phase: Clients reset the password with the help of clientid, password, and OTP.

4.2 Efficient Server Selection Using Modified Cuckoo Optimization Algorithm (MCOA)

The MCOA checks all available servers and calculates fitness for each server. This fitness is depending on the fitness of the virtual machines of particular server. So first calculate the fitness of virtual machine using Eq. (1).

$$M_{\text{fit}} = \frac{\sum_{j=1}^{s} \text{task}_j}{\left(\frac{\sum_{j=1}^{s} C_j \sum_{j=1}^{s} m_j}{u}\right) \times t} \tag{1}$$

where s is the number of virtual machines, C_j and M_j are the processor and the memory utilization, respectively, u is the number of memory units, task$_i$ represents the total number of task running in the jth virtual machine. M_{fit} is the fitness of virtual machine.

Modified cuckoo optimization algorithm [10]

1 Initialize cuckoo with eggs.
2 Lay eggs in different nest.
3 Some of eggs are detected and killed.
4 If population is less than maximum value, then check survival of eggs in nest and go to step 6.
5 If population is more than maximum value, then kill cuckoos in worst area and go to step 4.
6 If stop condition satisfied, then end.
7 If stop condition is not satisfied, then eggs grow.
8 Find nest with best survival rate.
9 Determine cuckoo societies.
10 Move all cuckoos toward best environment.
11 Determine egg laying radius for each cuckoo. Go to step 1.

4.3 Elliptic Curve Cryptography for Confidentiality

Elliptic curve cryptography [11] is the efficient method to encrypt and decrypt the data. Let p is prime number; F_p denote the field of integers modulo p. Given Eq. (2) described an elliptic curve C over F_p.

$$y^2 = x^3 + ax + b \tag{2}$$

where a, $b \in F_p$ satisfy $4a^3 + 27b^2 \neq 0 \,(\text{mod } p)$.

A pair (x, y) where $x, y \in F_p$ is a point on the curve if (x, y) satisfies Eq. (2). ∞ denoted the point at infinity. $C\left(F_p\right)$ denoted the set of all the points on C.

4.3.1 Key Generation

Key generation process is responsible to generate the public and secret key.

Elliptic curve Key Generation [11]

INPUT: Elliptic curve domain parameters (p, C, P, n).
OUTPUT: Public key K_p and Secret key K_s.
1. Select $K_s \in {}_R [1, n-1]$.
2. Compute $K_p = K_s P$.
3. Return (K_p, K_s).

4.3.2 Encryption

Encryption is the process which creates the cipher text from plain text.

Basic ElGamal Elliptic curve Encryption [11]

INPUT: Elliptic curve domain parameters (p, C, P, n), public key K_p, plaintext T.
OUTPUT: Cipher text (E_1, E_2).
1. Represent the message T as a point M in $C(F_p)$.
2. Select $k \in {}_R [1, n-1]$.
3. Compute $E_1 = kP$.
4. Compute $E_2 = M + kK_p$.
5. Return (E_1, E_2).

4.3.3 Decryption

Decryption is the process which creates plain text from cipher text.

Basic ElGamal Elliptic curve Decryption [11]

INPUT: Domain parameters (p, C, P, n), Secret key K_s, cipher text (E_1, E_2).
OUTPUT: Plaintext T.
1. Compute $M = E_2 - K_s E_1$, and extract T from M.
2. Return (T).

4.4 Optimize Cipher Text Using Covariance Matrix Adaptation Evolution Strategies (CMA-ES)

CMA-ES [13] is used to optimize the cipher text after encryption. It provides the optimal cipher text in terms of encryption quality, correlation coefficient, and differential attack. This evaluation-based scheme is suitable for such type of application. It is better than the genetic algorithm (GA) and particle swarm optimization (PSO). Here first we have generated different number of cipher text as for $E_2 = \{E_2^1, E_2^2, E_2^3, \ldots E_2^n\}$ same plain text. Now we have to find the best cipher text among generated cipher text, with the help of fitness function (O).

For calculating the fitness function, first we have to find encryption Quality (Q), Correlation coefficient (C_c), Information entropy factor (IE_f), and Avalanche effect (A_e)

Encryption Quality (Q)
Encryption quality of cipher text is depending on deviation. So first we calculate the D (bit deviation) and then find the $\overline{D}$ (average of deviation). Our aim is to maximize the encryption quality so here used objective function O_1 to find out the maximum value for Q.

$$D = |T - E_2^i|, \quad i = 1, 2, \ldots n \tag{3}$$

$$\overline{D} = \frac{1}{l}(D) \tag{4}$$

where l is the length of plain text.

$$Q = |D - \overline{D}| \tag{5}$$

Optimize function

$$O_1 = \max(Q) \tag{6}$$

Correlation coefficient (C_c)
C_c [14] parameter is used for measuring the correlation between the plain text and cipher text. It is calculated through Eq. (7). The value of C_c should be minimum for optimal result.

$$C_c = \frac{\text{cov}(T, E_2^n)}{\sqrt{\text{std}(T)}\sqrt{\text{std}(E_2^n)}} \tag{7}$$

The objective function O_2 is specified in Eq. (8), which find the min value of C_c.

$$O_2 = \min(C_c) \tag{8}$$

Information Entropy factor (IE_f)
IE_f is evaluated using Eq. (9) and then find its max value using Eq. (10).

$$IE_f = -\sum \{0 \leq i \leq (n-1)\} p(D_i)\log_2 p(D_i) \qquad (9)$$

where $p(D_i)$ represents probability of D_i.
Objective function O_3 is specified in Eq. (10)

$$O_3 = max\left(IE_f\right) \qquad (10)$$

Avalanche Effect (A_e)
It is calculated with the help of Eq. (11) and then finds its maximum value with the help of Eq. (12).

$$A_e = \frac{\text{Hamming Distance}}{\text{File Size}} \qquad (11)$$

Objective function

$$O_4 = max(A_e) \qquad (12)$$

For finding the fitness function, adding Eqs. (6), (8), (10), and (12)

$$O = O_1 + O_2 + O_3 + O_4 \qquad (13)$$

Fitness function (O)
Then overall objective function (fitness function) is calculated using Eq. (14) which is used to optimize the cipher text.

$$O = max\left(Q\right) + min(C_c) + Max\left(IE_f\right) + max\left(A_e\right) \qquad (14)$$

5 Conclusion

Cloud computing is an upcoming technology. Cloud architectures face problems regarding efficient computation as well as many security issues. Cloud is the collection of servers so selection of server for data storage is very important for efficient accessibility of cloud data. Biological searching algorithms are better to select the server for data storage. Many security models are available for ensuring the security of cloud data. This paper proposed a framework which integrates the efficient server selection, three-stage-authentication, ECC, and CMA-ES to achieve the security of data in cloud. Safe authentication prevents cloud data from illegal access and attacks.

MCOA is used to select the proficient server. For achieving the confidentiality, selection of efficient encryption technique ECC to encrypt the data and CMA-ES is used to find the optimal cipher text before stored the data on cloud. Both authentication and encryption ensure high privacy, confidentiality, and security of cloud data.

References

1. Habib SM, Hauke S, Ries S, Mühlhäuser M (2012) Trust as a facilitator in cloud computing: a survey. J Cloud Comput Adv Syst Appl 1:19
2. Singh A, Chatterjee K (2017) Cloud security issues and challenges: a survey. J Netw Comput Appl 79:88–115
3. Subashini S, Kavitha V (2011) A survey on security issues in service delivery models of cloud computing. J Netw Comput Appl 34(1):1–11
4. Younis YA, Kifayat K, Merabti M (2014) An access control model for cloud computing. J Inf Secur Appl 19(1):45–60
5. Madni SH, Latiff MS, Coulibaly Y (2016) Resource scheduling for infrastructure as a service (IaaS) in cloud computing: challenges and opportunities. J Netw Comput Appl 68:173–200
6. Kaur N, Chhabra A (2016) Analytical review of three latest nature inspired algorithms for scheduling in clouds. In: International conference on electrical, electronics, and optimization techniques 2016, IEEE, pp 3296–3300
7. Banyal RK, Jain P, Jain VK (2013) Multi-factor authentication framework for cloud computing. In: Fifth international conference on computational intelligence, modeling and simulation, IEEE, pp 105–110
8. Shobha Bindu C (2015) Click based graphical CAPTCHA to Thwart spyware attack. In: International advance computing conference (IACC), IEEE, pp 324–328
9. Jiang X, Ling J (2013) Simple and effective one-time password authentication scheme. In: 2nd international symposium on instrumentation and measurement, sensor network and automation, IEEE, pp 529–531
10. Mahmoudi S, Lotfi S (2015) Modified cuckoo optimization algorithm (MCOA) to solve graph coloring problem. Appl Soft Comput 33:48–64
11. Hankerson D, Menezes AJ, Vanstone S (2004) Guide to elliptic curve cryptography. Springer, Berlin, pp 32–40
12. Singh LD, Singh KM (2015) Implementation of text encryption using elliptic curve cryptography. Procedia Comput Sci 54:73–82
13. Kämpf JH, Robinson D (2009) A hybrid CMA-ES and HDE optimization algorithm with application to solar energy potential. Appl Soft Comput 9(2):738–745
14. Ahmed HEH, Kalash HM, Farag Allah OS (2007) Encryption efficiency analysis and security evaluation of RC6 block cipher for digital image. In: International conference on electrical engineering (ICEE), IEEE, pp 1–7

Dynamic Sentiment Analysis Using Multiple Machine Learning Algorithms: A Comparative Knowledge Methodology

Manmeet Kaur, Krishna Kant Agrawal and Deepak Arora

Abstract Human can easily understand or interpret the meaning of language. However, a machine has no natural language to deduce the hidden emotions. Without knowing the context of the word, it cannot simply infer whether a piece of text conveys joy, anger or frustration. Here, sentiment analysis came into picture. Sentiment analysis is the analysis of feelings, attitude and opinions of human emotions extracted from text. It uses natural language processing (NLP) for classifying the text into positive, negative or neutral category. Many businesses nowadays take feedback of the product from the customers to improve the quality or service of the product. Earlier feedbacks were taken by the call center executives but today a vast amount of data is available on the Internet. People share their views regarding products, services, people, etc. Sentiment analysis makes the task easier by extracting the relevant words from the sentences and classifying it in different categories. In this paper, we have described the essential steps used in the process of the sentiment analysis and few fields that work under its umbrella. A comparative analysis of machine learning algorithm like Naive Bayes, SVM, maximum entropy is done along with the few algorithms like artificial neural network and K-nearest neighbor, which can be used in sentiment analysis.

Keywords Machine learning · Support vector machine · Naive Bayes
Maximum entropy · Sentiment classification · Building resource
Transfer learning · Emotion detection · Chi square · Information gain

M. Kaur (✉) · K. K. Agrawal · D. Arora
Department of Computer Science and Engineering, Amity University, Noida,
Uttar Pradesh, India
e-mail: meetsoni2006@yahoo.co.in

K. K. Agrawal
e-mail: kkagrawal@outlook.com

D. Arora
e-mail: darora@lko.amity.edu

1 Introduction

The Internet today has become the vast sea of information and knowledge (a source of information). A large-scale enterprise to a small-scale firm uses the Internet for their business. People provide feedback of the products, rate them, share feelings about the politics and give review about the movie. It is providing a platform where everything is presented at one place. These data are useful for firms in order to improve the quality of the product to enhance their business. An automated machine learning system operates on this abundance data to perform sentiment analysis which eliminates the manual work of human by classifying the text in positive, negative or neutral category.

As shown in Fig. 1, sentiment analysis can be done using lexicon-based approach, machine learning approach and hybrid approach which have been described in next section in detail. The most used approach is machine learning in which the classification of text can be done using many supervised and unsupervised algorithms. Supervised learning makes use of the training documents or corpus, e.g., Naïve Bayes, SVM, maximum entropy. Unsupervised learning is used when no such labeled training documents are available, e.g., neighbor classification.

2 Background

A lot of work has been done in the field of sentiment analysis and its related field. Sentiment extraction can be done through different approaches, namely lexicon-based approach, machine learning approach, hybrid/combined approach [1].

2.1 Lexical-Based Approach

In the lexical-based approach, the text input is separated into several words, which is called tokenization. Once the word is tokenized, the numbers of positive, negative and neutral words are counted which are kept in the library of the words called Bag of Words (BoW). Further to classify the sentiment values of the document, lexicon is used.

In the process of classifying the sentiments, the system takes the count of the word which is being used in the statement and measures the value of goodness or badness of each of them, thus summing up and deciding on the sentiment of the text as a whole. This technique disregards the order of the words; e.g., it is not a good movie, is a negative sentence, but in such approach, this statement can be misinterpreted as a positive one. Lexical analysis has a drawback, with the exponential growth of the size of the dictionary (number of words); its performance (in terms of time complexity and accuracy) degrades drastically [1].

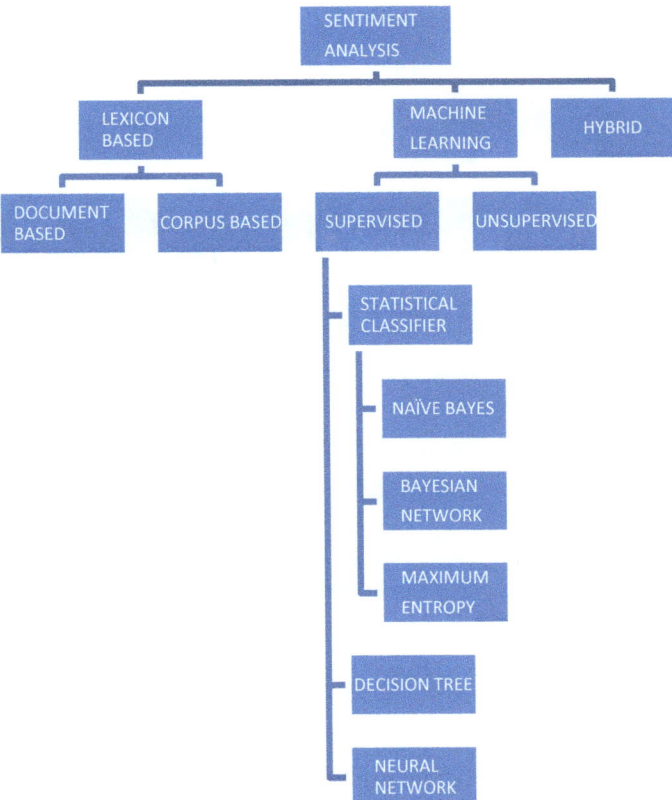

Fig. 1 Classification of sentiment analysis

2.2 Machine Learning Approach

Due to the adaptability and accuracy of this technique, it is one of the most prominent methods which is gaining attention in today's world. This classification of text can be done with the help of many supervised and unsupervised machine algorithms. Supervised learning makes use of the training documents or corpus, e.g., Naive Bayes, SVM, maximum entropy. Unsupervised learning is used when no such labeled training documents are available, e.g., neighbor classification. Artificial neural network-based algorithms are also used for the classification.

Fig. 2 Steps involved in sentiment analysis

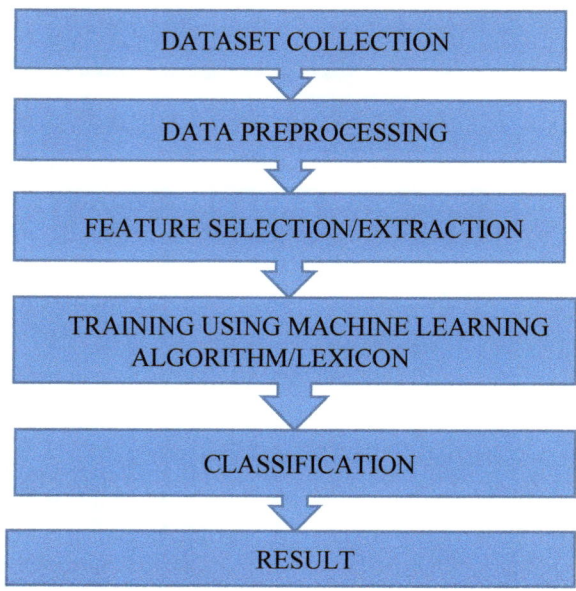

2.3 Hybrid Approach

With the advancement in the field of sentiment analysis, researchers were allured to find the hybrid approach which could exhibit the speed of lexical approach and the accuracy of a machine learning approach [1]. In a study, the authors collected the Chinese and English text from twitter [2]. Translate the Chinese text into English using Yahoo Babelfish and Google translator. They used lexicon approach (tokenization for creating Bag of Words) and unlabeled data for dividing the corpus of text into two categories—positive and negative. The training dataset obtained from it was further used for training purpose in Naive Bayes classifier for sentiment classification.

3 Methodology

In this section, the steps involved in the process of sentiment analysis have been described. Figure 2 depicts these steps as described below.

3.1 Dataset Collection

The first step is to collect the data from the various sources. The researchers have widely used the data from social networking and micro-blogging sites like Twitter as the dataset could be collected from it with the minimal supervision effort [2]. The data from IMDB (for movie reviews) and amazon.com (for customer's opinion/feedback) have been a great source of attraction for the researchers as these Web sites provide real-life datasets.

3.2 Data Preprocessing

This step is taken to clean the data and eliminate the noise. This step requires the elimination of stop words. Stop words are those words that do not carry any sentiments, e.g., of, have, been, to. Stemming is also done at this step in which the word is reduced to its root. For example, checking, checkers are the stemming words, and their root word is check. In this step, the data transformation can also be done if required.

3.3 Feature Selection/Extraction

This step in sentiment classification is used to extract features, i.e., text features. This could be finding adjectives, phrases which contribute to the importance of sentiments. Few features have been described in this section.

Text presence and frequency. These features could separate words or n-gram word and the frequency of occurrence of the words [3]. The selection of correct featured word is very important in sentiment analysis as the twisted words play a vital role in changing the essence of the statement. 'Not bad' holds a positive sentiment; if only bad is taken, then the sentence will be misinterpreted as the negative one. The use of n-gram helps in increasing the accuracy of the classification. The frequency of the words contributes to the importance of the feature [3].

Opinion words. These are the words that emphasize on individual opinions, whether bad or good, like or dislike. Some phrases implicitly express the opinion as well, for example cost a pretty penny.

Negation. The presence of negative words may change the essence of the statement. For example, it was not a pleasant day, holds a negative sentiment, although pleasant holds a positive sentiment.

Feature selection is a crucial step in sentiment analysis. Few algorithms used by researchers for feature selection are described below.

Information Gain. Information gain is used as an attribute selection method [4]. The motive is to select the attribute that is most useful in classification of examples. It measures how well a given tuple separates the training example according to its target classification. The attribute with the highest information gain, also known as entropy, is selected for the classification. It is used in decision making to select an attribute which is used in the classification of node of the tree. It assures the purity of the partition.

The information needed to classify the attribute is given by

$$\inf(D) = - \sum_{i=1}^{n} p_i \log p_i \tag{1}$$

where $\inf(D)$ is the actual information, pi is the probability of the attribute.

$$\inf{}_a(D) = h \sum_{j=1}^{v} \frac{|D_j|}{|D|} * \inf o \left(D_j \right) \tag{2}$$

$$\mathrm{Gain}(a) = \inf o(D) - \inf o_a(D) \tag{3}$$

$\inf_a(D)$ is the expected information needed to classify a tuple, and $\mathrm{Gain}(a)$ is the information gain.

Chi Square (λ^2). Chi square finds the deviation between the observed count and expected count.

$$\lambda^2(a, b) = \frac{N(wz - yx)^2}{(w + y)(x + z)(w + x)(y + z)} \tag{4}$$

where w, x, y, z represent the frequencies that tell the absence or presence of the feature in a sample dataset [5]. W represents count of sample in which feature a and class b occurred together. It actually represents 'a' as the feature and 'b' as the class.

Other methods used for the feature selection are Gini index that measures the impurity of a set of data tuple, hidden Markov model (HMM), PMI, LDA, etc.

After the feature extraction, sentiment analysis is implemented through lexicon-based approach or machine learning approach which classifies the text into positive, negative or neutral category. In the next section, various machine learning algorithms have been discussed.

4 Classification Methods

In this section, the various supervised and unsupervised machine learning algorithms used for the purpose of classification of text in sentiment analysis have been discussed.

4.1 Naive Bayes

Bayesian classifiers are statistical classifiers. They can predict the class membership probability that the tuple belongs to the specific class. It assumes that the effect of values of an attribute of given class is independent of the values of other attributes.

This is called conditional independence. This works on the Bag of Word feature extraction method and makes independent assumption about the occurrence of words. For a given feature 'f' and class category 'c,' the conditional probability that 'f' belongs to 'c' can be formulated as:

$$P(c/f) = \frac{p(f/c) * p(c)}{p(f)} \qquad (5)$$

$P(f)$ is the prior probability of the feature that it has occurred [3]. $P(f/c)$ is the prior probability that the given feature is being classified as the class label. $P(c/f)$ is the posterior probability that the given feature belongs to a particular class. Given Naive assumption, the features will be considered as independent of each other. The equation can be rewritten as

$$P(c/f) = \frac{p(c) * p(f1/c) * p(f2/c) * \cdots * p(fn/c)}{p(f)} \qquad (6)$$

The main advantage of the Naive Bayes classifier is that it requires small amount of training data that calculates the parameters which is used for prediction [6].

4.2 Support Vector Machine (SVM)

Support vector machine is a non-probabilistic binary linear classifier that works on both linear and nonlinear data. Each review is represented in vectorized form which shows a data point in space [6]. It finds the best hyperplane, $\underset{w}{\rightarrow}$, to separate the textual data vector having maximum marginal distance. Once the training of the model is done, the testing data are mapped into some space, which is used to predict on which side of the hyperplane the data fall [6].

Let $c_i \in \{1, -1\}$ represent the positive and negative classes, respectively, for a document $\underset{d_i}{\rightarrow}$, and the equation for $\underset{w}{\rightarrow}$ is given by

$$\vec{w} = \sum_i \alpha_i c_i d_i, \quad \alpha_i \geq 0 \qquad (7)$$

All the $\underset{d_i}{\rightarrow}$ such that $\alpha_i > 0$ are said to be support vectors [7].

Tripathi et al. used labeled polarity movie dataset having 1000 negative and 1000 positive reviews [6]. Data cleaning is done to extract the relevant features. The vectorization technique was used to convert the word data into numeric format. A matrix was created using vectorization, where each row and column represented an individual review and a feature, respectively. K-fold cross-validation technique was applied on the matrix to select the training and testing dataset for each fold [6]. Table 1 represents the comparison between the existing literatures.

	Pang and Lee	Read	Tripathi et al.
Naive Bayes	0.864	0.789	0.895
SVM	0.8615	0.815	0.940

Table 1 Comparison based on the existing literature

Pang and Lee as well as Tripathi et al. in their paper used ten fold cross-validation to perform classification, while Read used only three folds [6]. It showed that the higher number of folds results in more generalized result and SVM gives more accurate result than Naive Bayes.

4.3 Maximum Entropy Classifier

Maximum entropy is a common technique for the estimation of the probability distributions from data [8]. The principal behind maximum entropy says that the correct distribution is when the constraints set by the 'evidence' are met even when the entropy/uncertainty is maximized. Constraints are expected values of the features.

Maximum entropy classifier is a machine learning technique which is based on empirical data. Nigam et al. and Berger et al. showed that in many cases it outperforms Naive Bayes classification [9]. Raychaudhari et al. also found that maximum entropy worked better than Naive Bayes and nearest neighbor classification for their classification [10]. Unlike Naive Bayes, it does not make independent assumption about the occurrence of words. The mathematical formula for entropy is given by

$$H = \sum p(c, d) \log p(c, d) \tag{8}$$

where $p(c, d)$ is the probability that the document 'd' belongs to the specific category 'c.'

Kamal et al. showed in their research that maximum entropy outperforms the regular Naive Bayes as the classification error is reduced by more than 40% [8]. They experimented on webKB dataset. But in comparison with scaled Naive Bayes, the results were mixed. In scaled Naive Bayes, each document is scaled such that it contains constant number of word count.

In [8], the author implemented maximum entropy technique in three datasets out of which two performed better than Naive Bayes. Basic maximum entropy suffers from over-fitting. They used Gaussian prior to reduce the over-fitting problem resulted in a better performance. They showed that the feature selection is an essential factor for maximum entropy.

Fig. 3 Multilayer
feed-forward neural
network [4]

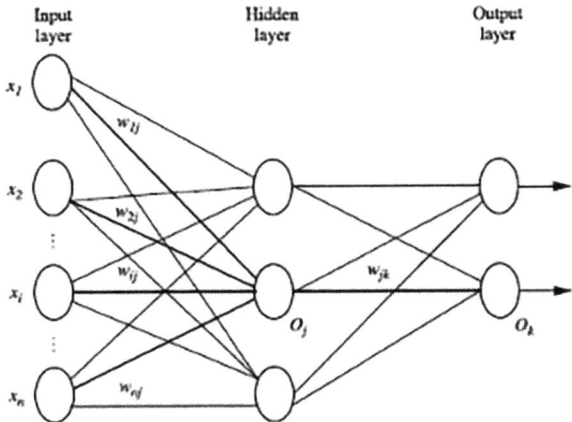

4.4 Artificial Neural Network

Artificial neural network is composed of artificial neurons, which are its basic unit. It consists of input layer, hidden layer and output layer [4]. This network consists of connected input/output units. There is a weight associated with every neuron. The units associated with an input layer are called input units. The weighted sums of the given inputs are added to the activation function that gives an output which acts as an input to the hidden layer. This is a feed-forward network as the output to one layer acts as an input to the other. This network has poor interpretability (Fig. 3).

4.5 Backpropagation Neural Network (BNN)

Backpropagation neural network is a feed-forward neural network in which the error rate is reduced through backpropagation as the actual output is matched with the expected output and then error is propagated backward by updating the weights and the bias (which helps in varying the activity) [4].

The word frequencies to the jth document act as an input to the neuron. The weight W is associated with each and every neuron. The linear function associated with it is given as $I = W.X_j$. For the binary classification, the sign predicts the class label. For the nonlinear classification, multilayer neural network is used. The complexity of the training dataset increases as the size of the middle layer increases because the error is backpropagated over the different layers. Smaller hidden layer tends to produce better classification of the class label.

There is a comparison between SVM and neural network [11]. They showed that ANN outperformed SVM when the experiment is conducted on the movie reviews. There is a limitation associated with both the techniques like the computational cost of ANN at the training time and SVM at the running time.

Steps for the backpropagation algorithm:

S1: Take the input from the input node.

S2: First input from the input node is equal to the output for this node which acts as the input to this network.

$$O_i = I_i \tag{9}$$

S3: Input to the hidden layer or output layer is the weighted sum from the previous nodes added by bias (where B_j is the bias), i.e.,

$$I_j = \sum_i W_{ij} O_i + B_j \tag{10}$$

S4: Output of the hidden layer or output layer can be found using activation function that could be sigmoid or logistic function.

$$O_j = \frac{1}{1 + e^{-I_j}} \tag{11}$$

S5: The errors are backpropagated for all the unit j in the output layer.

$$E_k = O_j \left(1 - O_j\right) \left(Y - O_j\right) \tag{12}$$

where Y is the expected output

S6: The errors are backpropagated from last hidden layer to the first hidden layer for all the unit j.

$$E_k = O_j \left(1 - O_j\right) \sum_k E_k w_{jk} \tag{13}$$

S7: weights are updated.

$\Delta w_{ij} = E_j O_i(l)$, l is the learning rate, $0 < l < 1$, and l is used to avoid local minimum and encourages to reach global maxima in a decision surface.

$$W_{ij} = \Delta W_{ij} + W_{ij} \tag{14}$$

S8: For updating bias (Fig. 4)

$$\Delta B_j = E_j(l) \tag{15}$$

$$B_j = \Delta B_j + B_j \tag{16}$$

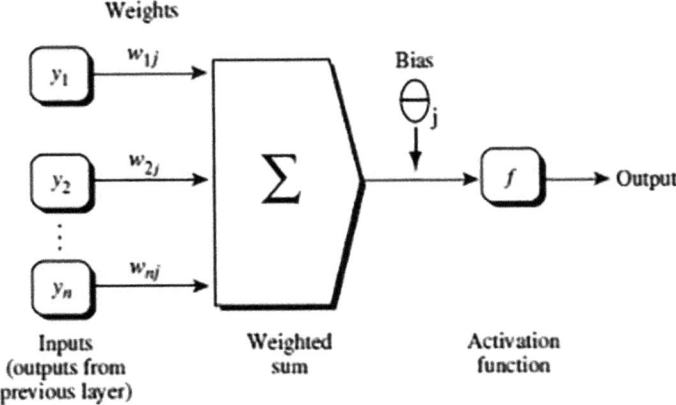

Fig. 4 Weighted sum of the output of the previous layer is added with a bias which is followed by an activation function that gives the output [4]

4.6 Semi- and Unsupervised Learning

The idea behind the classification of text is to divide it into categories. This requires labeled training data or document in supervised leaning. However, it is often difficult to get the labeled data, but the unlabeled data or document can be obtained easily. Here, the unsupervised learning can be used that does not require labeled training data. Zhou et al. worked on a strategy at the level of the features instead of instances by providing weak supervision [12].

The prior information was incorporated into sentiment classifier model, obtained as an initial classifier, extracted from an existing sentiment lexicon. They worked on the data obtained from IMDB for movie reviews and amazon.com. Their work identified that the polarity of the words may vary from domain to domain. They showed for any text classification, the approach used by them obtained better performance than weakly supervised learning techniques if the little bit relevant prior knowledge is available.

4.7 PCA

Principal component analysis (PCA) is an unsupervised learning technique. Its motive is to find principal component (PC). The data are expressed in a manner that highlights their differences and similarities. This technique emphasizes on the variations and develops strong patterns in dataset. It helps to eliminate dimensions. PCA takes a dataset with a lot of dimensionality and flattens it to 2D or 3D so we can look at it. It tries to find a meaningful way to flatten the data by focusing on things that are different from words or sentences in a document. PCA finds direction in which it is orthogonal and with maximum variance.

4.8   K-Nearest Neighbor (KNN)

K-nearest neighbor classifiers are the lazy learners. When an unknown tuple is given, then KNN searches the pattern space which defines the closeness to the unknown attribute. The closeness can be found using Euclidean distance in terms of distance metrics. K number of tuples are chosen, and the other attributes are computed in terms of the distance. K defines the number of separate clusters. The unknown tuple is assumed to belong to the cluster which has the minimum distance with that cluster.

5   Applications

Sentiment analysis can be implemented in various fields that have attracted many researchers. Some of these fields have been discussed in this section.

5.1   Emotion Detection

Natural language processing can be used to implement sentiment analysis which helps in obtaining opinion about an entity. Opinion defines an attitude towards an object. Sentiments define the feelings. Emotion reflects attitude. Emotions can be joy, agony, sadness, frustration, disgust, anticipation and surprise. Sentiment analysis defines positive, negative or neutral opinion, whereas emotion detection defines various emotions. According to Plutchik, emotion detection is considered as sentiment analysis task [13].

Lu and Lin proposed an approach for detecting emotions on different events based on Web-based text mining [3]. This approach was formed on the probability distribution to find common mutual actions between the object and subject of an event.

Balahur et al. used lexicon-based approach and ML both [3]. They used the approach of common sense. They said that emotions cannot always be detected explicitly, i.e., by defining a sad word to express the emotion of grief or a happy word to express joy. SVM and SVM-SO approaches were used to achieve the goal. They showed that the use of EmotiNet (corpus storing the knowledge based on common sense) gives better results than supervised learning techniques and corpus-based approach.

5.2 Building Resource

Building resource focuses on creating dictionary, corpora and lexica in which expressions giving opinions are annotated in accordance with their polarity.

Robaldo et al. introduced building corpus in which they used opinion mining—ML- and XML-based tagging of textual expressions that convey the relevant opinion [14]. A standard methodology was used by them that annotated relevant statements in the text which was independent of any application domain. Then, domain-specific adaptation was considered that depended on the ontology used which is domain dependent. They used dataset on restaurant reviews, and query-oriented extraction process was used. The result showed that the annotation scheme was able to cover the large complexity along with the preservation of good agreement between different people.

5.3 Transfer Learning

Transfer learning obtains the knowledge from the auxiliary domain which is used to improve the learning phase in the target domain, for instance a search in Hindi to English. It is a new cross-domain learning method as it shows the several aspects of domain differences.

In SA, it can be used to build bridge between two domains as it transfers classification of sentiments from one domain to another.

6 Conclusion and Future Scope

After analyzing the various articles, it is observed that the feature selection method is a crucial step in sentiment analysis. Machine learning is the most popular approach used for the classification of sentiment analysis as the lexical approach suffers in terms of performance when the size of the dictionary increases. Many researchers have proved that SVM works better than Naive Bayes or other approaches for the sentiment classification. Yet every technique has few drawbacks and cannot resolve all the challenges.

Artificial neural network is a significant method for the classification which is gaining popularity due to its adaptive nature of adjusting themselves according to the data without explicitly specifying the functional form of the underlying project. The related fields, emotion detection, transfer learning and building resources are the emerging fields of research.

One of the biggest challenges related to the sentiment analysis is language. Interest in other languages than English is increasing and it requires the ability to analyze the emotions independent of languages. Google translator is the best tool for translating a language into English.

For the future, we would like to work on artificial neural network for the sentiment classification and analysis on dataset in other languages like Hindi or Punjabi along with the English language. Work in the field of emotion detection is also an open area of research as the study suggests that a positive word can be a sarcasm which could lead to unsuccessful classification.

References

1. Thakkar H, Patel D (2015) Approaches for sentiment analysis on twitter: a state-of-art study, 1–8
2. Pak A, Paroubek P, Paris-sud D, Limsi-cnrs L, Cedex FO (2010) Twitter based system: using twitter for disambiguating sentiment ambiguous adjectives. Comput Linguist, 436–439
3. Medhat W, Hassan A, Korashy H (2014) Sentiment analysis algorithms and applications: a survey. Ain Shams Eng J 5:1093–1113
4. Jiawei H, Kamber M (2001) Data mining: concepts and techniques
5. Shahana PH, Omman B (2015) Evaluation of features on sentimental analysis. Procedia Comput Sci 46:1585–1592
6. Tripathy A, Agrawal A, Rath SK (2015) Classification of sentimental reviews using machine learning techniques. Procedia Comput Sci 821–829
7. Pang B, Lee L, Vaithyanathan S (2002) Thumbs up: sentiment classification using machine learning techniques. Proc Conf Empir Methods Nat Lang Process, 79–86
8. Nigam K, Lafferty J, McCallum A (1999) Using maximum entropy for text classification. In: IJCAI-99 workshop on machine learning for information filtering, pp 61–67
9. Mehra N, Khandelwal S, Patel P (2002) Sentiment identification using maximum entropy analysis of movie reviews
10. Raychaudhuri S, Chang JT, Sutphin PD, Altman RB (2002) Associating genes with gene ontology codes using a maximum entropy analysis of biomedical literature. Genome Res 12:203–214
11. Moraes R, Valiati JF, Gavião Neto WP (2013) Document-level sentiment classification: an empirical comparison between SVM and ANN
12. Cao Q, Duan W, Gan Q (2011) Exploring determinants of voting for the "helpfulness" of online user reviews: a text mining approach. Decis Support Syst 50:511–521
13. Plutchik R (1980) A general psychoevolutionary theory of emotion. Emot Theor Res Exp 1:3–33
14. Robaldo L, Di Caro L (2013) OpinionMining-ML. Comput Stand Interfaces 35:454–469

Sentiment Analysis Using Tuned Ensemble Machine Learning Approach

Pradeep Singh

Abstract With the recent emergence of Web-based applications and use of social networking sites, number of people are eager in expressing their views and opinions online. The sentimental analysis also referred to as opinion mining aims at processing user reviews (about products, movies, services, books, places, etc.). These reviews are often unstructured and need processing to evolve into the productive knowledge. Majority of the sentiment analysis works on the classification of opinion polarity with the use of simple classifiers. Handling diverse data distribution is one of the major issues that simple classifiers suffer. To cope up with the issue in this paper, we utilized the ensemble learners on the polarity prediction of the movie reviews. The proposed work processes the review data through some elementary steps that are conducted for the feature extraction in sentiment analysis. In addition to the feature extraction, we further perform the feature selection for the sake of dimensionality reduction. However, in contrast to the conventional simple learner, we applied the ensemble learner in the proposed model and evaluated its performance. To compare the ensemble model competence, we conducted the experiment on both individual as well as ensemble learner (random forest, AdaBoost, extra trees) and computed classification measures on both the model. IMDB dataset is used, and the polarity of a review, i.e., whether it is positive or negative, is predicted. With an extensive experimentation, it is found that results of ensemble classifiers are outperforming than individual learner in the classification of sentiment polarity.

Keywords Sentiment analysis · Ensemble learner · Tuning of parameter

1 Introduction

In our daily life, the opinions of customers and users of a product have a great influence on our decision making. These decisions may range from buying electronic

P. Singh (✉)
Department of Computer Science & Engineering, National Institute of Technology, Raipur, India
e-mail: psingh.cs@nitrr.ac.in

© Springer Nature Singapore Pte Ltd. 2018
M. L. Kolhe et al. (eds.), *Advances in Data and Information Sciences*, Lecture Notes in Networks and Systems 38, https://doi.org/10.1007/978-981-10-8360-0_27

appliances or jewelry to taking review about the schools for children. Before the advent of Internet, opinions on products and services are taken from friends, relatives, or consumer reports. Now in the Internet era, it is much easier to collect diverse opinions from various people across the world. The review sites (CNET, Epinions, etc.), e-commerce sites (Flipkart, Amazon, Snapdeal, eBay, etc.), online opinion sites (TripAdvisor, Rotten Tomatoes), and social networking sites (Facebook, Twitter, etc.) are referred to get opinion about how a particular product or service is provided to them. Similarly, most of the organizations use opinion polls, surveys, and social media as a mode to obtain feedback on their products and services [1]. Sentimental analysis is the branch of text mining which processes these reviews computationally for identifying and categorizing opinions, sentiments, attitudes, subjectivity, views, evaluations, appraisal, emotions, etc., stated in a textual form as positive, negative, or neutral.

In sentimental analysis, classification is done according to different criteria such as polarity of the sentiment (negative, positive, or neutral), whether the opinion is in support or opposition of a service, number of pros and cons in the reviews, whether the user agrees or disagrees with some particular topic. According to [2] sentiment, classification is of three levels. Document level: The sentiment is evaluated by taking the whole document as one information unit. Sentence level: The sentiment is evaluated by taking each sentence as an individual unit. Aspect level: According to [3], in this level, the sentiment is evaluated by taking the polarity of each aspect of the review such as screenplay, acting skills, and direction for a movie. The sentiment analysis of movie reviews can be at document or aspect level [4].

Machine learning algorithms play a critical role in the document-level polarity prediction. The choice of supervised learning algorithm in the classification problem is cumbersome due to the wide availability of the candidate. Single classifiers are producing the good classification rate, but the presence of differences in the data distribution between train and test instances makes the simple learner to perform poorly. Thus, ensemble learner has the ability to handle the data distribution efficiently and with superior performance delivered by multi-classifier model in the other application makes an appropriate choice to be elected as alternative for the single classifier. This paper aims at classifying the document-level sentiments using machine learning algorithms and compares the performances of simple and ensemble learners. In addition to it, we also aid the joint contribution of unigram and bigrams features along with the parts of speech (POS tagging) for feature extraction, whereas feature selection algorithm is used for the efficient prediction of polarity with minimum data handling complexity. We considered random forest, multinomial naive Bayes, Bernoulli naive Bayes, SGD, SVC, and NuSVC for the sake of performance comparison between simple and ensemble learner. The contribution of the paper can be summarized as follows:

- Use of unigrams features leads to false analysis. Hence, we focus on the combination of uni- and bigrams which increase the efficiency.
- Use of feature selection reduces the total number of features (words), thereby decreasing the time for overall computation.

- Use of POS tagging eliminates many useless pronouns, propositions.

The remaining part of the paper is organized as follows: Sect. 2 presents contextual information related to the past work done. Section 3 presents the proposed sentiment analysis method covering common problems listed during the study of related work. Comprehensive discussion on the supervised learning on both simple and ensemble learners is presented in Sect. 4. In the last Sect. 5, we conclude the overall performance of the learners in sentiment analysis.

2 Related Work

Pang et al. considered aspect of sentiment analysis using categorization with positive and negative sentiments [5]. They used different machine learning algorithms (classifiers) like support vector machine, Naïve Bayes. They classified it using unigram features, bigram features, and by combining both unigram and bigram features. To realize the algorithm, they use bag-of-words (BOW) in their algorithms and found SVM with good classification rate.

Salvetti [6] discussed an overall opinion polarity (OvOP). Here hypernym given by word net and parts of speech tagging acts as lexical filter. The results from word net are less accurate than POS filter. Their work has shown good result in Web data. Mullen [7] applied SVM where values are given to selected words and are pooled to make a model for the classification. Features which are close to the topic are allocated with higher values. They gave comparison of their approach and hand annotation. Their approach gave better results.

Matsumoto [8] used syntactic relation among words in document-level sentiment analysis. The frequent word subsequence and dependency sub trees are extracted from the sentences and used as input features for support vector machines (SVM) machine learning algorithm. They performed their experiment on IMDB and polarity dataset. Liu [9] proposed multi-label classification. They used 11 methods compared on 2 micro-blog dataset and 8 evaluation matrices. Lin et al. [10] performed an empirical study of sentiment categorization on Chinese hotel review. A Chinese corpus, MioChnCorp, with a million Chinese hotel reviews is collected. A word2vec model is trained using MioChnCrop to represent words and phrases in Chinese hotel domain. Their experimental results indicate that the more data produces the better performance. They also used word embeddings which represent each comment as input in different machine learning methods like SVM, Logistic Regression, Convolutional Neural Network (CNN) and ensemble methods for sentiment classification.

The literature review identifies the vague issues that remained untouched during the problem solving. From the review, we find the following issues:

- Most of the work done in the field of sentimental analysis is done considering unigram approach which may not include all the features accurately.

- Much work has been done on the sentiment analysis, but they have used a lot of features for the feature extraction part. A lot of time is wasted during the processing of the whole review data. If there is a big dataset then it wastes a lot of time.
- Majority of the model uses single classifiers which performs poor in the case of diverse data.

In this paper, we aim at providing the solutions for the issues that were found during the survey. The objective is achieved by collecting sufficiently large amount of data consisting of reviews, preprocessing it applying feature selection in order to extract the features with high frequencies and reducing large number of words to a limited features and passing these inputs to the ensemble classifiers; additionally, we also performed the simulation with the simple classifiers for the sake of performance comparison.

3 Methodology

The process followed in this paper is shown in Fig. 1. In this model, we acquired the textual datasets and applied preprocessing and POS tagging on them. This preprocessed datasets is divided into train and test data (if not given explicitly) by tenfold cross-validation, for the further use of train data in learner. Later, we input the test data to predict the sentiment. The original sentiment of the reviews is compared with the obtained sentiment to calculate accuracy, precision, recall, and f-measure.

To evaluate the performance of simple and ensemble learner on sentiment analysis, we carried the elementary steps that are followed in the opinion mining process. Firstly, IMDB reviews used in this project were acquired from [11]. This dataset consists of 25,000 reviews. Apart from the text data, the dataset has numeric, acronym, and HTML tag contents in it. Thus, in order to have good classification it is necessary to eliminate these entries from the dataset.

To remove such entries, we apply preprocessing on the entire dataset. This preprocessing involves removal of acronyms, numeric letters, and HTML tag. In addition to this, we also eliminate the stop words (words which do not contribute to the sentiment analysis).

An acronym is an abbreviation, and it is generally used in content published on the Internet. Users are often drawn toward brief words, and using acronyms is one way of ensuring that the sentence still grabs the attention of the reader despite the fact that it is short. For example, DND stands for 'Do Not disturb' while OMG stands for 'Oh My God.'

Reviews contain numeric characters which do not affect the sentiment of words. Hence they are removed. Stop words such as 'comma,' 'full stop,' parenthesis, question marks, exclamations, and special characters such as *, @, $, # are eliminated. Also the reviews that contain references to contain links are useless. Hence they are eliminated.

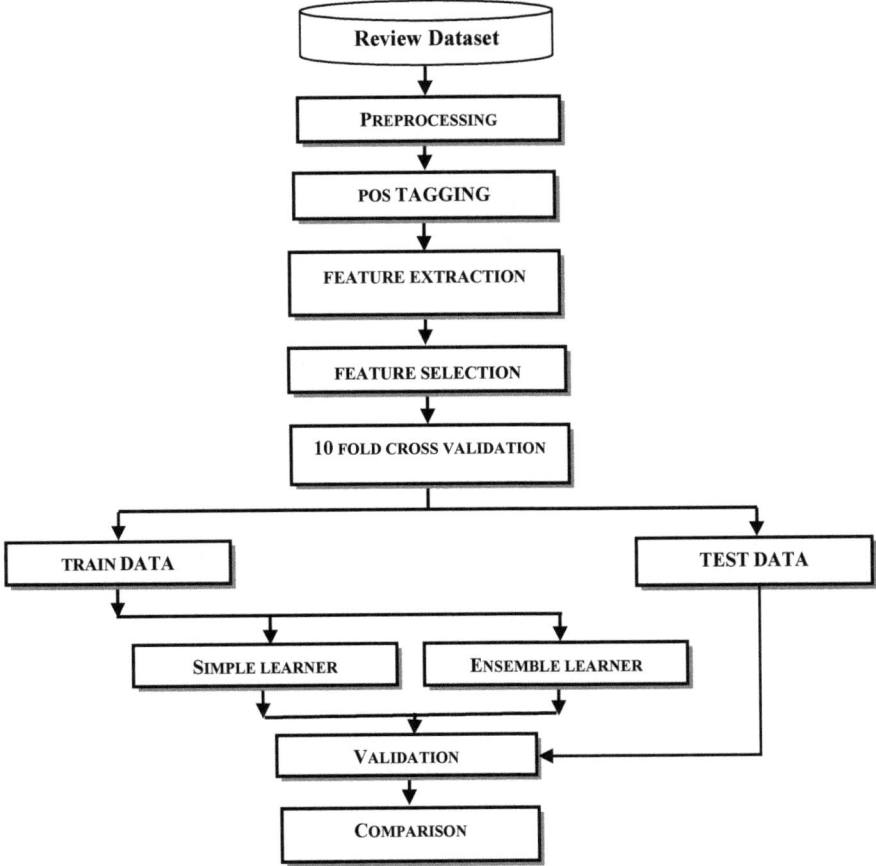

Fig. 1 Process flow diagram

After preprocessing, parts of speech (POS) tagging is performed in the processed data. The review carries the word, and tagging assigns parts of speech to each word in the dataset, i.e., whether the word is a noun, pronoun, verb, adjective, or adverb. POS tagging can be used as feature reduction by selecting only few elements in parts of speech; i.e., pronouns and prepositions can be eliminated from the data which reduce the number of features. In our work, we used a combination of verbs, nouns, adjectives, and adverbs and eliminated the remaining features.

Feature extraction is an attribute or dimensionality reduction process [12]. It is used to extract a subset of features from the original feature set by means of some functional mapping retaining as much info in the data as possible as mentioned in [13]. Feature extraction is used to transform the actual attributes. The transformed features are linear combinations from the original attributes. Models constructed based on extracted features tend to be of higher quality, because the information is described by less, more meaningful attributes. Feature extraction techniques used in our project are:

Bigrams: Bigrams are nothing but a sequence of two adjacent words in the sentence. Examples of bigrams are 'not good,' 'very bad.' Bigrams are very useful in assigning correct polarity as the string of 2 words can increase the efficiency. Bigram is an N-gram for $N = 2$.

Term frequency-inverse document frequency (TF-IDF): Term frequency is the amount of times a specific word or term appears in the text. Inverse document frequency measures the existence of a particular word in all documents. According to [14], the values of TF-IDF are directly proportional to the term frequency; i.e., it increases as the frequency of a word in the document increases.

Count vectorizer: It implements both counting of occurrence and tokenization in a single class. It converts the entire text documents into a sparse matrix representation.

4 Supervised Learning Algorithms

We employed nine classifiers from the scikit-learn package [15] from Python, four from the simple learner, and four from the ensemble learners. The hyper-parameters of classifiers are tuned using randomized parameter optimization.

Classifier	Type	Description
Random forest	Ensemble	It uses a huge number of individual, unpruned decision trees
AdaBoost [16]	Ensemble	Amount of focus is quantified by a weight that is allocated to every pattern in the exercising set
Extra trees [17]	Ensemble	Randomizing tree building in the context of numerical input features, where the choice of the optimal cut-point is responsible for a large proportion of the variance of the induced tree
Gradient boosting [18]	Ensemble	'boosting' many weak predictive versions into a strong one, available as ensemble of weak types
Support vector machines [19]	Simple	Kernel-based method uses hyperplane that separates the classes and has the largest distance between border line data points
Naïve Bayes variants, Multinomial Naive Bayes (MNB) Bernoulli Naive Bayes [20]	Simple	The probabilistic model of Naïve Bayes is originated from Bayes theorem
Stochastic gradient descent [21]	Simple	Stochastic approximation of gradient descent optimization approach which is used to minimize an objective function
Logistic regression	Simple	Logistic regression is a linear model for classification also known as maximum entropy classification (MaxEnt) or the log-linear classifier

5 Experimental Results

When we use a classifier model, we always need to find the exactness of that model as the result obtained forecasts from all expected results. This is called classifier accuracy. When we have to choose whether it is a sufficient model to take care of, accuracy is not the only metric for assessing the viability of a classifier. Two other important measurements are precision and recall. A confusion matrix is a summary of prediction results of a classification problem (Table 1).

A false positive error, or in short false positive, commonly called a 'false alarm,' is a result that specifies a given condition has been satisfied; when it actually has not been satisfied, i.e., erroneously a positive effect has been assumed. A false negative error, or in short false negative, is where a test result indicates that a condition failed; while it actually was successful. True positives are relevant items that we correctly identified as relevant. True negatives are irrelevant items that we correctly identified as irrelevant. A confusion matrix C is such that $C_{(i,j)}$ is equal to the number of observations known to be in group i but predicted to be in group j. A confusion matrix is used to describe the performance of the classifier.

Accuracy is how close a measured value is to the actual (true) value. It is the proportion of instances whose class the classifier can correctly predict. It can be calculated as shown in Eq. 1.

$$\text{Accuracy} = \frac{T_p + T_n}{\text{Total number of samples}} \tag{1}$$

where T_P denotes the number of true positives and T_n is the number of true negatives.

Precision measures exactness of a classifier. High precision indicates that there are less number of false positives; likewise, a lower precision indicates more number of false positives. Precision is defined as the ratio of number of true positives over the number of true positives plus the number of false positives [22]. Its formula is shown in Eq. 2.

$$P = \frac{T_p}{T_p + F_p} \tag{2}$$

where F_p denotes the number of false positives.

Recall is used to measure completeness, or sensitivity, of a classifier. Increasing the value of recall often decreases precision because it gets increasingly harder to be more precise as sample space increases. Recall is defined as the number of true

Table 1 Confusion matrix

Actual class	Predicted class	
	Yes	No
Yes	T_P	F_N
No	F_P	T_N

Table 2 Performance metrics—random forest classifier

Feature type	IMDB dataset			
	Accuracy	Precision	Recall	F-measure
UNIGRAM	84.7	83.48	85.56	84.51
UNI + BI	84.73	83.72	85.45	84.57
Parameter tuning (UNI)	85.33	84.65	85.81	85.23
Parameter tuning (UNI + BI)	85.64	85.15	85.99	85.57

positives over the number of true positives plus the number of false negatives. Its formula is shown in Eq. 3.

$$R = \frac{T_p}{T_p + F_n} \tag{3}$$

Precision and recall can be combined to produce a single metric known as F-measure. It is weighted harmonic mean of precision and recall. Its equation is shown in Eq. 4.

$$F1 = 2\frac{P \times R}{P + R} \tag{4}$$

where P is the precision and R is the recall.

We compare the performance of the classifiers that we used based on their precision, recall, F-measure, accuracy, and confusion matrices.

We train each of the classifiers using the two datasets individually. First, we test all the classifiers on each training set one at a time and then we test it on the test set. Table 2 summarizes the results of all the experiments performed on IMDB dataset.

The results of different approaches on the IMDB dataset are shown in Table 3. Initially, the analysis is done only by considering the unigram words and applying the random forest machine learning algorithm. We used noun, verb, adjective, and adverb from our data and removed all the unnecessary words. On applying single machine learning algorithms (Table 3) on the IMDB dataset, we observed good results in NuSVC and stochastic gradient descent method. On applying on ensemble classifiers (Table 4), extra trees classifier using (uni + bi) has performed outstanding.

After applying the parameter tuning on the random forest classifier (Table 2), the accuracy of the sentiment prediction has increased by 1%. The parameters used in our process are n_estimators = 100, max_features='sqrt', oob_score='true', n_jobs = −1, random_state = 50.

Table 3 Performance metrics—single classifiers

Classifier and feature type	IMDB dataset			
	Accuracy	Precision	Recall	*F*-measure
NuSVC UNIGRAM	87.452	89.032	86.305	87.647
NuSVC (UNI + BI)	87.342	88.804	86.281	87.524
SGD UNI	90.532	87.096	93.522	90.195
SGD UNI + BI	90.530	87.096	93.519	90.193
MNB UNI	83.968	81.304	85.879	83.529
MNB UNI + BI	84.728	83.648	85.495	84.561
BNB UNI	84.192	81.7	85.985	83.788
BNB UNI + BI	85.022	84.296	85.538	84.912
LR UNI	84.369	82.485	85.715	84.069
LR UNI + BI	85.084	84.4	85.571	84.981

Table 4 Performance metrics—ensemble classifiers

Classifier and feature type	IMDB dataset			
	Accuracy	Precision	Recall	*F*-measure
AdaBoost (UNI)	83.468	84.176	81.395	82.762
AdaBoost (UNI + BI)	83.8	84.28	81.86	83.05
ExtraTrees (UNI)	86.128	84.528	87.32	85.90
ExtraTrees (UNI + BI)	86.496	85.11	87.535	86.306
RandomForest (UNI)	84.7	83.48	85.568	84.511
RandomForest (UNI + BI)	84.736	83.72	85.456	84.579
Gradient boosting (UNI)	81.096	86.336	78.146	82.037
Gradient boosting (UNI + BI)	81.14	86.208	78.274	82.049

6 Conclusions

In this paper, intensive experiments were performed to predict movie reviews using different supervised machine learning algorithms like Naïve Bayes (NB), stochastic gradient descent (SGD), support vector machines (SVM), random forest, AdaBoost, extra trees, and gradient boosting. We applied both unigram, unigram + bigram approach. The learner performed better when using unigram + bigram approach than the unigram feature. Parameter tuning has been applied to improve the accuracies.

Processing of Twitter reviews may have some issues because they contain emojis and smileys (they hold important information whether it is a positive tweet or negative tweet) which are not processed in our approach. Some words like 'greatttt', 'fineee' are also processed using the stemmer because those features should not be missed. The accuracy of the prediction may increase with various preprocessing techniques and machine learning algorithms. Taking the above limitations into consideration, further work can be performed in order to improve the accuracy of sentiment prediction.

References

1. Fernández-gavilanes M, Álvarez-lópez T, Juncal-martínez J, Costa-montenegro E, González-castaño FJ (2016) Unsupervised method for sentiment analysis in online texts, vol 58, pp 57–75
2. Medhat, W., Hassan, A., Korashy H (2014) Sentiment analysis algorithms and applications: a survey. Ain Shams Eng 5(4):1093–1113
3. Parkhe V (2014) Aspect based sentiment analysis of movie reviews
4. Singh VK, Piryani R, Uddin A (2013) Sentiment analysis of movie reviews, pp 712–717
5. Pang B, Lee L, Vaithyanathan S (2002) Thumbs up: sentiment classification using machine learning techniques. Proc Conf Empir Methods Nat Lang Process, 79–86
6. Salvetti F, Lewis S, Reichenbach C (2004) Automatic opinion polarity classification of movie. Color Res Linguist 17(1):2
7. Mullen T, Collier N (2004) Sentiment analysis using support vector machines with diverse information sources. Conf Empir Methods Nat Lang Process, 412–418
8. Matsumoto S, Takamura H, Okumura M (2005) Sentiment classification using word subsequences and dependency sub-trees. In: Proceedings of 9th Pacific-Asia conference advances in knowledge discovery and data mining, vol 059, pp 301–311
9. Liu SM, Chen J-H (2015) A multi-label classification based approach for sentiment classification. Expert Syst Appl 42(3):1083–1093
10. Lin Y, Lei H, Wu J, Li X (2015) An empirical study on sentiment classification of Chinese review using word embedding. In: 29th Pacific Asia conference on language information and computation, pp 258–266
11. http://ai.stanford.edu/~amaas/data/sentiment/
12. https://docs.oracle.com/database/121/DMCON/feature_extr.htm#DMCON268
13. Pechenizkiy M, Puuronen S, Tsymbal A (2001) Feature extraction for classification in the data mining process PCA-based feature extraction feature extraction for a classifier and dynamic integration of classifiers. Int J 10:271–278
14. Tripathy A, Agrawal A, Rath SK (2016) Classification of sentiment reviews using n-gram machine learning approach. Expert Syst Appl 57:117–126

15. Pedregosa F (2011) Scikit-learn: machine learning in python. J Mach Learn Res 12:2825–2830
16. McCallum A, Nigam K (1998) A comparison of event models for Naive Bayes text classification. AAAI/ICML-98 work learning for text categorization, pp 41–48
17. Kibriya AM (2004) Multinomial Naive Bayes for text categorization revisited. Adv Artif Intell, 488–499
18. Mason L, Baxter J, Bartlett P, Frean M (1999) Boosting algorithms as gradient descent. Nips, 512–518
19. Fradkin D, Muchnik I (2006) Support vector machines for classification. Discret Methods Epidemiol 70:13–20
20. http://sebastianraschka.com/Articles/2014_naive_bayes_1.html
21. https://books.google.co.in/books?id=48u5BQAAQBAJ&pg=PA369&lpg=PA369&dq=Stochastic+Gradient+Descent
22. http://machinelearningmastery.com/classification-accuracy-is-not-enough-more-performance-measures-you-can-use/